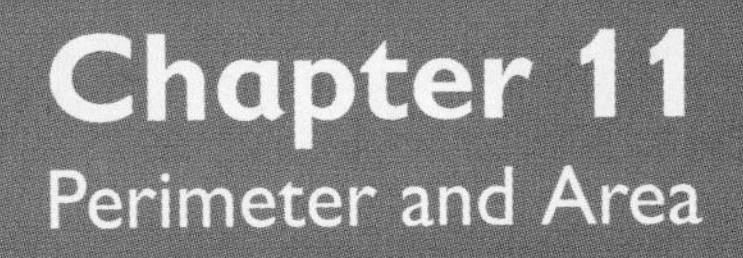

Chapter 11
Perimeter and Area

Houghton Mifflin Harcourt

Printed in the U.S.A.

ISBN 978-0-544-34217-0

20 0928 20

4500800120 C D E F G

Dear Students and Families,

Welcome to **Go Math!**, Grade 3! In this exciting mathematics program, there are hands-on activities to do and real-world problems to solve. Best of all, you will write your ideas and answers right in your book. In **Go Math!**, writing and drawing on the pages helps you think deeply about what you are learning, and you will really understand math!

By the way, all of the pages in your **Go Math!** book are made using recycled paper. We wanted you to know that you can Go Green with **Go Math!**

Sincerely,

The Authors

Made in the United States
Text printed on 100% recycled paper

GO MATH!

Authors

Juli K. Dixon, Ph.D.
Professor, Mathematics Education
University of Central Florida
Orlando, Florida

Edward B. Burger, Ph.D.
President, Southwestern University
Georgetown, Texas

Steven J. Leinwand
Principal Research Analyst
American Institutes for Research (AIR)
Washington, D.C.

Matthew R. Larson, Ph.D.
K-12 Curriculum Specialist for Mathematics
Lincoln Public Schools
Lincoln, Nebraska

Martha E. Sandoval-Martinez
Math Instructor
El Camino College
Torrance, California

Contributor

Rena Petrello
Professor, Mathematics
Moorpark College
Moorpark, California

English Language Learners Consultant

Elizabeth Jiménez
CEO, GEMAS Consulting
Professional Expert on English Learner Education
Bilingual Education and Dual Language
Pomona, California

Measurement

Critical Area Developing understanding of the structure of rectangular arrays and of area

Critical Area

Perimeter and Area 623

COMMON CORE STATE STANDARDS

3.MD Measurement & Data

Cluster C Geometric measurement: understand concepts of area and relate area to multiplication and to addition.
3.MD.C.5, 3.MD.C.5a, 3.MD.C.5b, 3.MD.C.6, 3.MD.C.7, 3.MD.C.7a, 3.MD.C.7b, 3.MD.C.7c, 3.MD.C.7d

Cluster D Geometric measurement: recognize perimeter.
3.MD.D.8

✓ Show What You Know 623
Vocabulary Builder 624
Chapter Vocabulary Cards
Vocabulary Game 624A
1 Investigate • Model Perimeter 625
Practice and Homework
2 Find Perimeter 631
Practice and Homework
3 Algebra • Find Unknown Side Lengths 637
Practice and Homework
4 Understand Area 643
Practice and Homework
5 Measure Area 649
Practice and Homework
6 Use Area Models 655
Practice and Homework
✓ Mid-Chapter Checkpoint 661

GO DIGITAL

Go online! Your math lessons are interactive. Use *i*Tools, Animated Math Models, the Multimedia eGlossary, and more.

Chapter 11 Overview

In this chapter, you will explore and discover answers to the following **Essential Questions**:

- How can you solve problems involving perimeter and area?
- How can you find perimeter?
- How can you find area?
- What might you need to estimate or measure perimeter and area?

7 Problem Solving • Area of Rectangles **663**
Practice and Homework

8 Area of Combined Rectangles **669**
Practice and Homework

9 Same Perimeter, Different Areas **675**
Practice and Homework

10 Same Area, Different Perimeters **681**
Practice and Homework

✓ Chapter 11 Review/Test **687**

CRITICAL AREA REVIEW PROJECT ZOO ANIMAL HABITATS: *www.thinkcentral.com*

Practice and Homework

Lesson Check and Spiral Review in every lesson

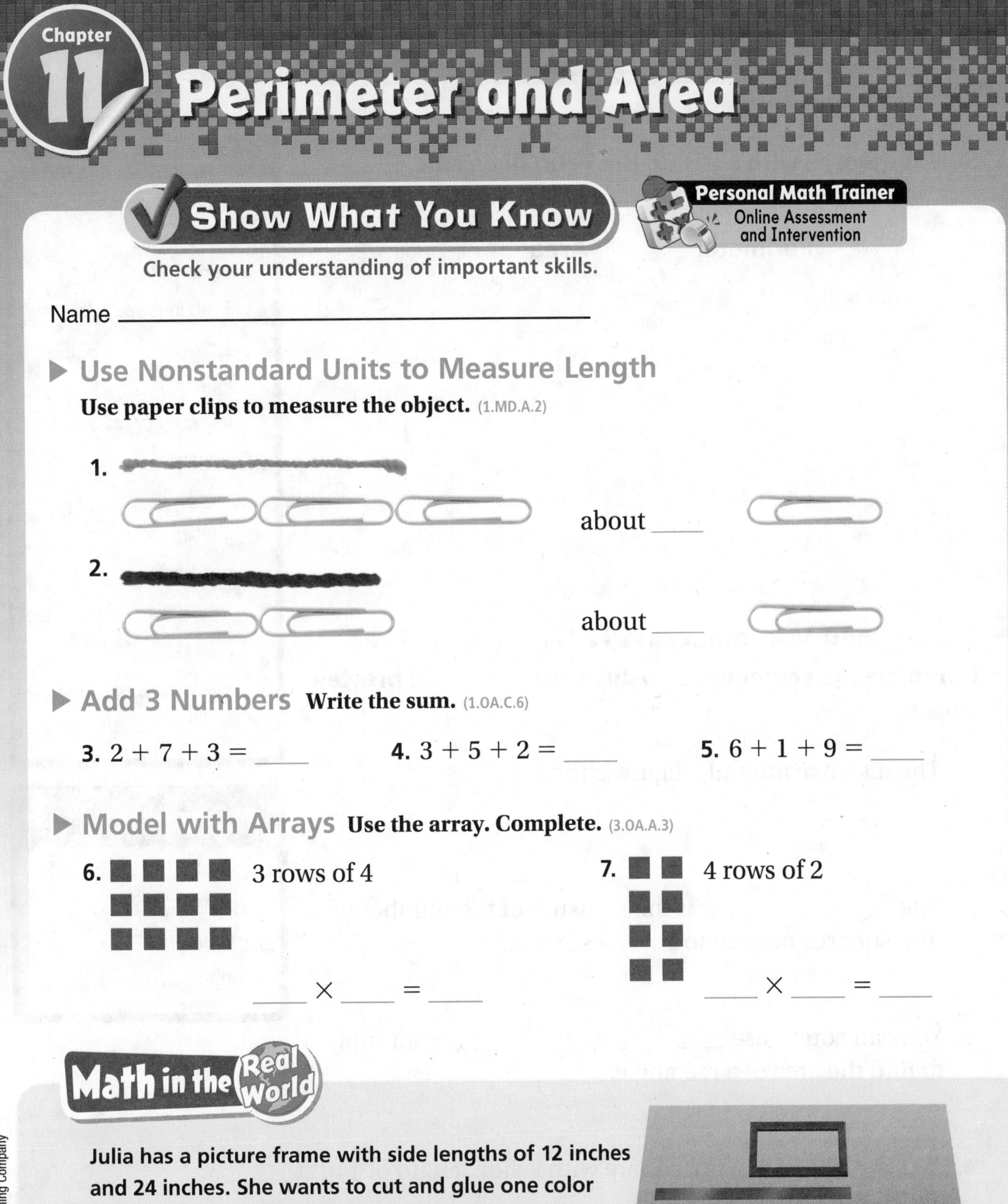

Chapter 11 Perimeter and Area

Show What You Know

Personal Math Trainer
Online Assessment and Intervention

Check your understanding of important skills.

Name ______________________________

Use Nonstandard Units to Measure Length

Use paper clips to measure the object. (1.MD.A.2)

1. about ____

2. about ____

Add 3 Numbers

Write the sum. (1.OA.C.6)

3. $2 + 7 + 3 =$ ____

4. $3 + 5 + 2 =$ ____

5. $6 + 1 + 9 =$ ____

Model with Arrays

Use the array. Complete. (3.OA.A.3)

6. 3 rows of 4

____ × ____ = ____

7. 4 rows of 2

____ × ____ = ____

Math in the Real World

Julia has a picture frame with side lengths of 12 inches and 24 inches. She wants to cut and glue one color of ribbon that will fit exactly around the edge. The green ribbon is 72 inches long. The red ribbon is 48 inches long. Find which ribbon she should use to glue around the picture frame.

Vocabulary Builder

▶ Visualize It

Sort the words with a ✓ into the Venn diagram.

Perimeter | Area

Review Words

- addition
- array
- centimeter (cm)
- Distributive Property
- foot (ft)
- inch (in.)
- inverse operations
- ✓ length
- meter (m)
- multiplication
- pattern
- rectangle
- repeated addition
- ✓ unit
- ✓ width

Preview Words

- area
- perimeter
- ✓ square unit
- ✓ unit square

▶ Understand Vocabulary

Complete the sentences by using the review and preview words.

1. The distance around a figure is the _______________.
2. The _______________ is the measure of the number of unit squares needed to cover a surface.
3. You can count, use _______________, or multiply to find the area of a rectangle.
4. A _______________ is a square with a side length of 1 unit and is used to measure area.
5. The _______________ shows that you can break apart a rectangle into smaller rectangles and add the area of each smaller rectangle to find the total area.

GO DIGITAL
- Interactive Student Edition
- Multimedia eGlossary

Chapter 11 Vocabulary

A metric unit used to measure length or distance
100 centimeters = 1 meter

0 1 2 3 4 5
centimeters

The measure of the number of unit squares needed to cover a surface

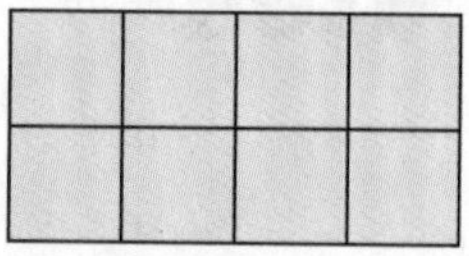

Area = 8 square units

The measurement of the distance between two points

Opposite operations, or operations that undo one another, such as addition and subtraction or multiplication and division

Examples: $16 + 8 = 24$; $24 - 8 = 16$
$4 \times 3 = 12$; $12 \div 4 = 3$

A quadrilateral with two pairs of parallel sides, two pairs of sides of equal length, and four right angles

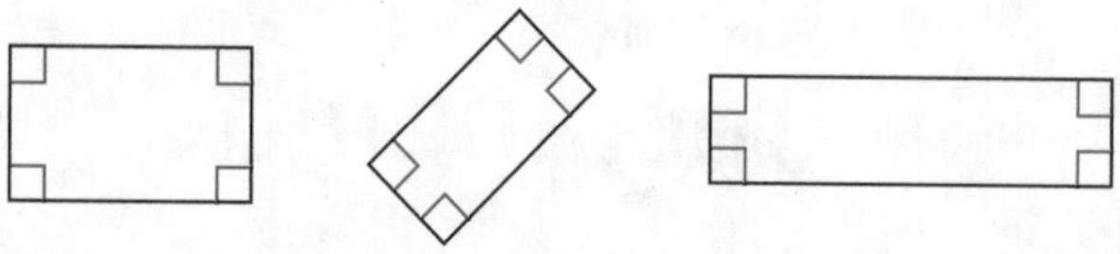

The distance around a figure

Example: The perimeter of this rectangle is 20 inches.

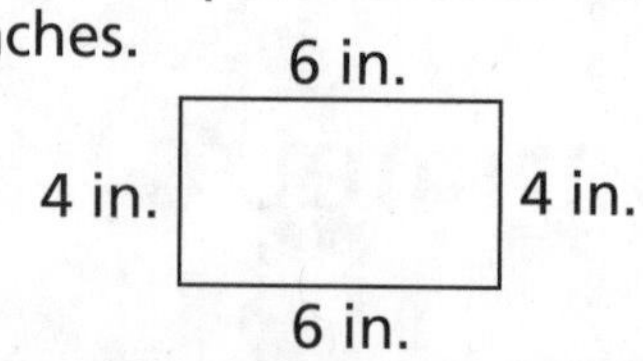

6 in.
4 in. 4 in.
6 in.

A square with a side length of 1 unit, used to measure area

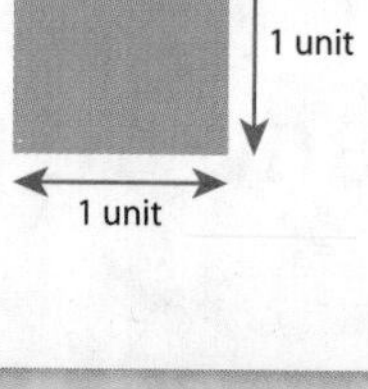

1 unit
1 unit

A unit used to measure area such as square foot, square meter, and so on

Game

Picture It

Word Box
- area
- centimeter (cm)
- inverse operations
- length
- perimeter
- rectangle
- square unit
- unit square

For 3 to 4 players

Materials
- timer
- sketch pad

How to Play
1. Take turns to play.
2. To take a turn, choose a word from the Word Box, but do not say the word aloud.
3. Set the timer for 1 minute.
4. Draw pictures and numbers to give clues about the word.
5. The first player to guess the word before time runs out gets 1 point. If that player can use the word in a sentence, he or she gets 1 more point. Then that player gets a turn choosing a word.
6. The first player to score 10 points wins.

The Write Way

Reflect

Choose one idea. Write about it.

- Define perimeter in your own words.
- Write two things you know about area.
- Explain how two rectangles can have the same area, but different perimeter. Give an example.

Name ______________________________

Model Perimeter

Essential Question How can you find perimeter?

Common Core **Measurement and Data—3.MD.D.8**

MATHEMATICAL PRACTICES
MP1, MP3, MP6, MP7

Investigate

Hands On

Perimeter is the distance around a figure.

Materials ■ geoboard ■ rubber bands

You can find the perimeter of a rectangle on a geoboard or on dot paper by counting the number of units on each side.

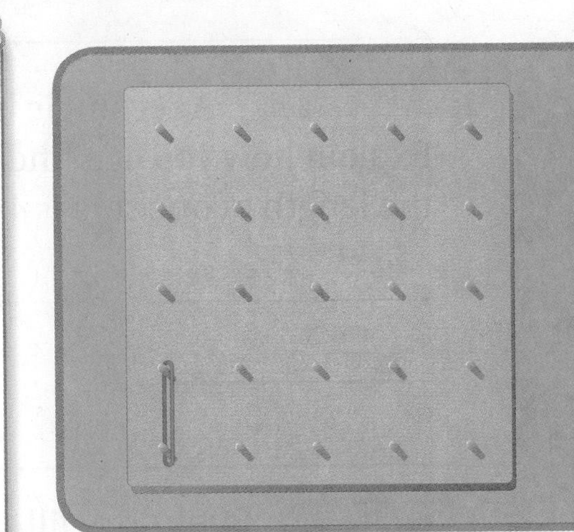

A. Make a rectangle on the geoboard that is 3 units on two sides and 2 units on the other two sides.

B. Draw your rectangle on this dot paper.

← 1 Unit

C. Write the length next to each side of your rectangle.

D. Add the number of units on each side.

_____ + _____ + _____ + _____ = _____

E. So, the perimeter of the rectangle is _____ units.

- How would the perimeter of the rectangle change if the length of two of the sides was 4 units instead of 3 units?

____

____

Draw Conclusions

1. Describe how you would find the perimeter of a rectangle that is 5 units wide and 6 units long.

____

2. THINKSMARTER A rectangle has two pairs of sides of equal length. Explain how you can find the unknown length of two sides when the length of one side is 4 units, and the perimeter is 14 units.

____

____

____

3. MATHEMATICAL PRACTICE 1 **Evaluate** Jill says that finding the perimeter of a figure with all sides of equal length is easier than finding the perimeter of other figures. Do you agree? Explain.

____

____

Make Connections

Math Talk

MATHEMATICAL PRACTICES 3

Apply If a rectangle has a perimeter of 12 units, how many units wide and how many units long could it be?

You can also use grid paper to find the perimeter of figures by counting the number of units on each side.

Start at the arrow and trace the perimeter. Begin counting with 1. Continue counting each unit around the figure until you have counted each unit.

A

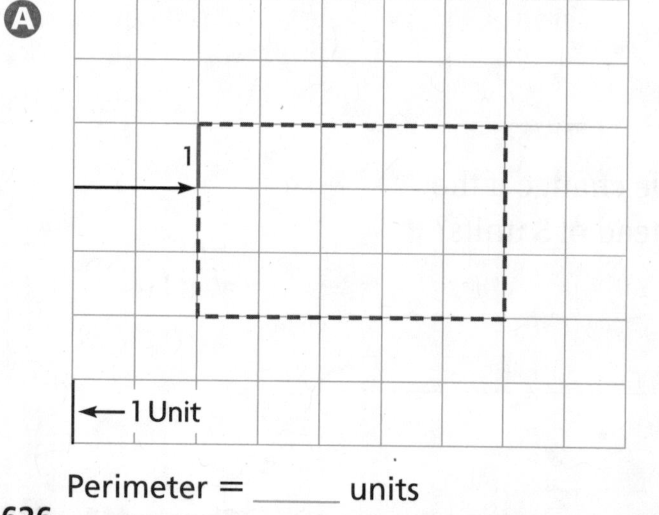

Perimeter = _____ units

B

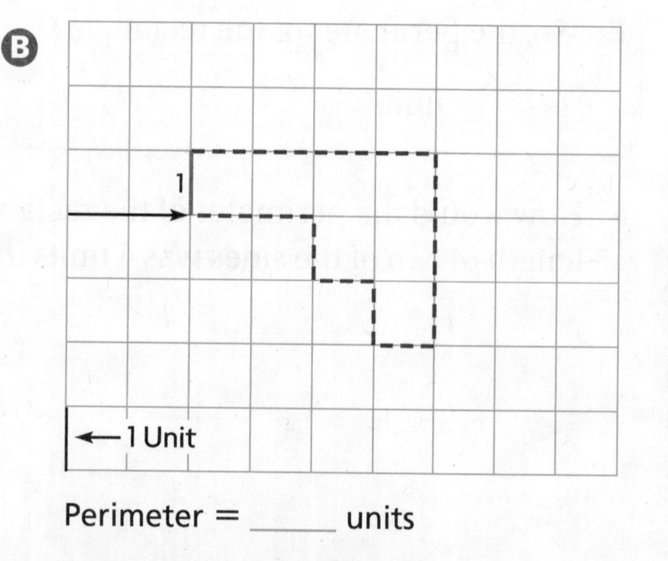

Perimeter = _____ units

Name ______________________

Find the perimeter of the figure. Each unit is 1 centimeter.

1.

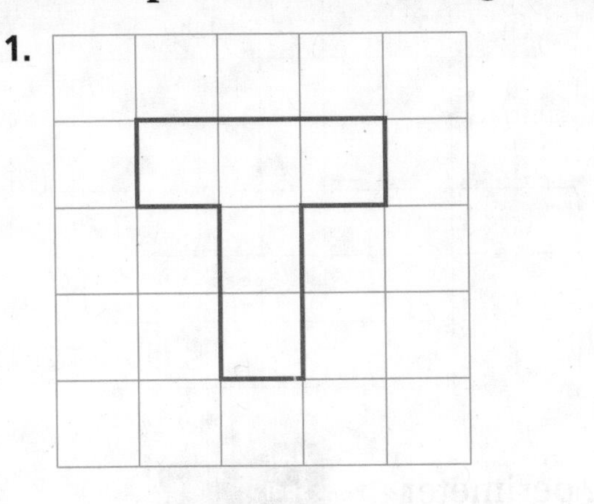

____ centimeters

2.

____ centimeters

3.

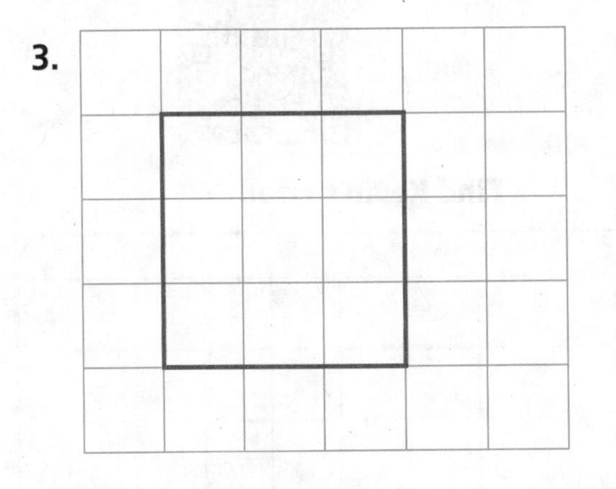

____ centimeters

4.

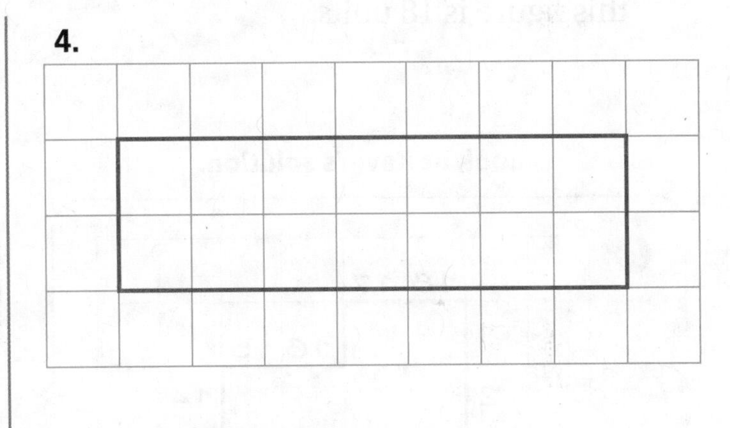

____ centimeters

Find the perimeter.

5. A figure with four sides that measure 4 centimeters, 6 centimeters, 5 centimeters, and 1 centimeter

____ centimeters

6. A figure with two sides that measure 10 inches, one side that measures 8 inches, and one side that measures 4 inches

____ inches

Problem Solving • Applications

Real World

7. MATHEMATICAL PRACTICE 6 **Explain** how to find the length of each side of a triangle with sides of equal length and a perimeter of 27 inches.

____

____

8. THINK SMARTER Luisa drew a rectangle with a perimeter of 18 centimeters. Select the rectangles that Luisa could have drawn. Mark all that apply. Use the grid to help you.

- (A) 9 centimeters long and 2 centimeters wide
- (B) 6 centimeters long and 3 centimeters wide
- (C) 4 centimeters long and 4 centimeters wide
- (D) 5 centimeters long and 4 centimeters wide
- (E) 7 centimeters long and 2 centimeters wide

9. THINK SMARTER **What's the Error?** Kevin is solving perimeter problems. He counts the units and says that the perimeter of this figure is 18 units.

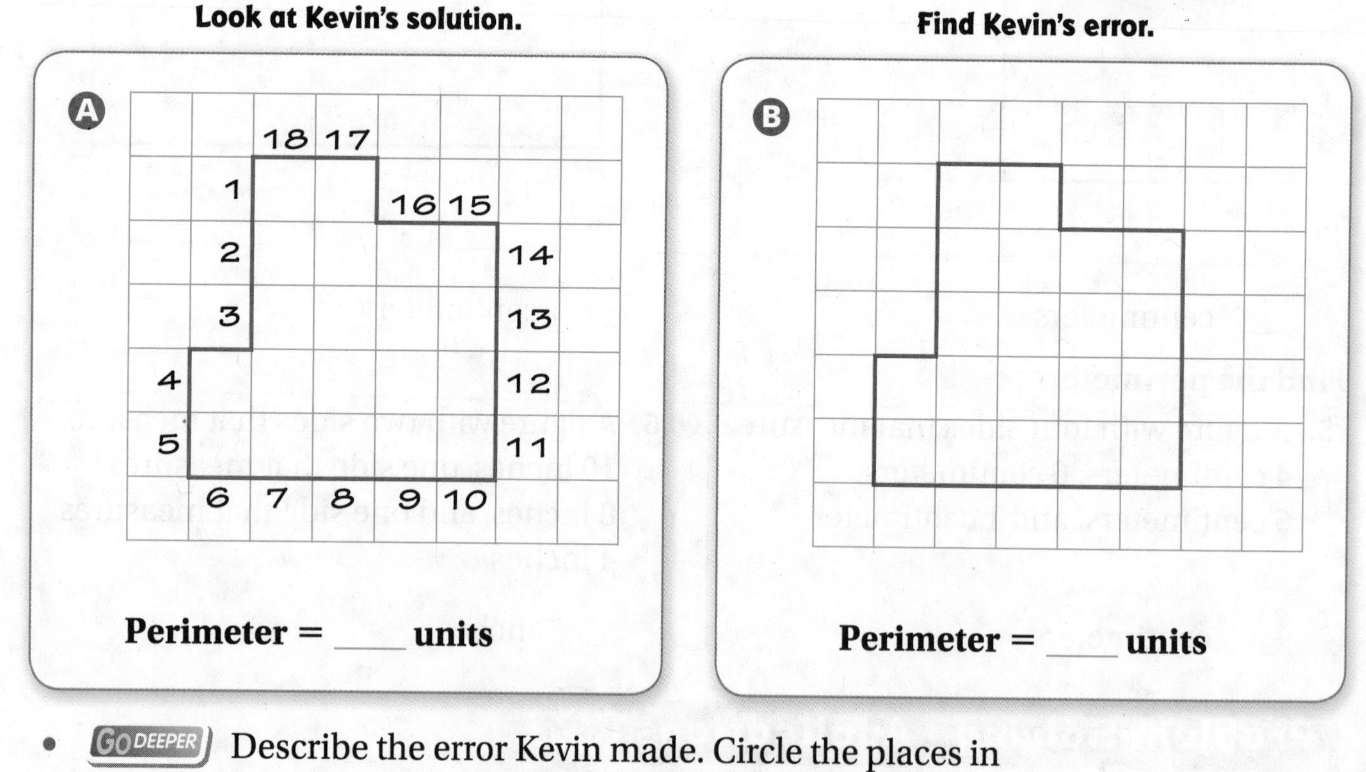

- GO DEEPER Describe the error Kevin made. Circle the places in the drawing of Kevin's solution where he made an error.

____

____

Name ______________________

Model Perimeter

COMMON CORE STANDARD—3.MD.D.8
Geometric measurement: recognize perimeter as an attribute of plane figures and distinguish between linear and area measures.

Find the perimeter of the figure. Each unit is 1 centimeter.

1.

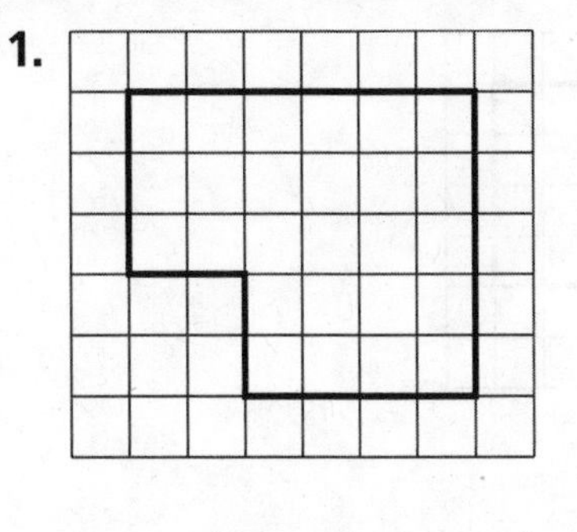

___22___ centimeters

2.

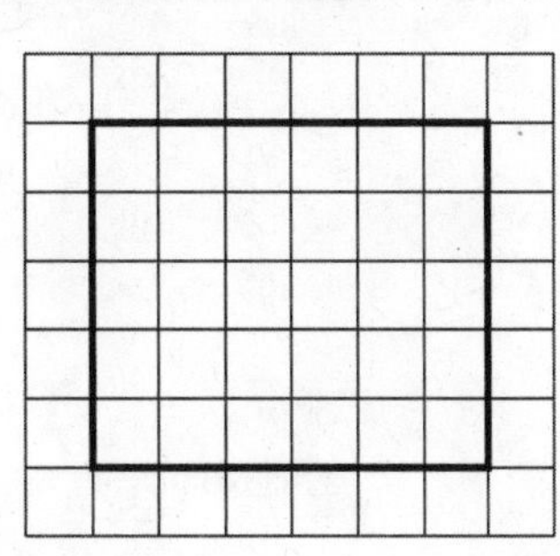

__________ centimeters

Use the drawing for 3–4. Each unit is 1 centimeter.

Patrick's Figure

Jillian's Figure

3. What is the perimeter of Patrick's figure?

____

4. How much greater is the perimeter of Jillian's shape than the perimeter of Patrick's figure?

____

5. WRITE *Math* Draw a rectangle and another figure that is not a rectangle by tracing lines on grid paper. Describe how to find the perimeter of both figures.

____

____

____

Lesson Check (3.MD.D.8)

1. Find the perimeter of the figure. Each unit is 1 centimeter.

2. Find the perimeter of the figure. Each unit is 1 centimeter.

Spiral Review (3.NF.A.3d, 3.MD.A.1, 3.MD.A.2)

3. Order the fractions from least to greatest.

$$\frac{2}{4}, \frac{2}{3}, \frac{2}{6}$$

4. Kasey's school starts at the time shown on the clock. What time does Kasey's school start?

5. Compare. Write <, >, or =.

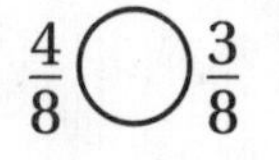

6. Aiden wants to find the mass of a bowling ball. Which unit should he use?

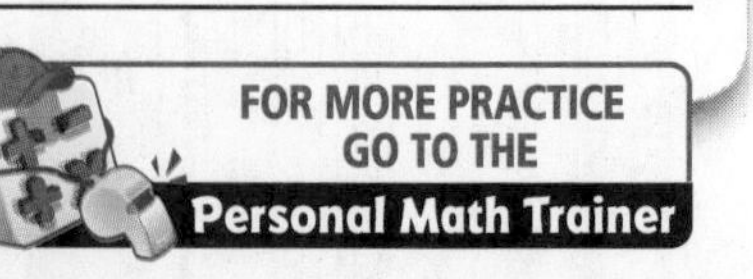

Name ______________________

Find Perimeter

Essential Question How can you measure perimeter?

Common Core **Measurement and Data—3.MD.D.8** *Also 3.NBT.A.2, 3.MD.B.4*

MATHEMATICAL PRACTICES
MP1, MP2, MP4, MP5

You can estimate and measure perimeter in standard units, such as inches and centimeters.

Unlock the Problem Real World

Hands On

Find the perimeter of the cover of a notebook.

Activity **Materials** ■ inch ruler

STEP 1 Estimate the perimeter of a notebook in inches. Record your estimate. _____ inches

STEP 2 Use an inch ruler to measure the length of each side of the notebook to the nearest inch.

STEP 3 Record and add the lengths of the sides measured to the nearest inch.

_____ + _____ + _____ + _____ = _____

So, the perimeter of the notebook cover measured to the nearest inch is _____ inches.

Math Talk MATHEMATICAL PRACTICES 1

Evaluate How does your estimate compare with your measurement?

Try This! Find the perimeter.

Use an inch ruler to find the length of each side.

Add the lengths of the sides:

_____ + _____ + _____ + _____ = _____

The perimeter is _____ inches.

Use a centimeter ruler to find the length of each side.

Add the lengths of the sides:

_____ + _____ + _____ + _____ = _____

The perimeter is _____ centimeters.

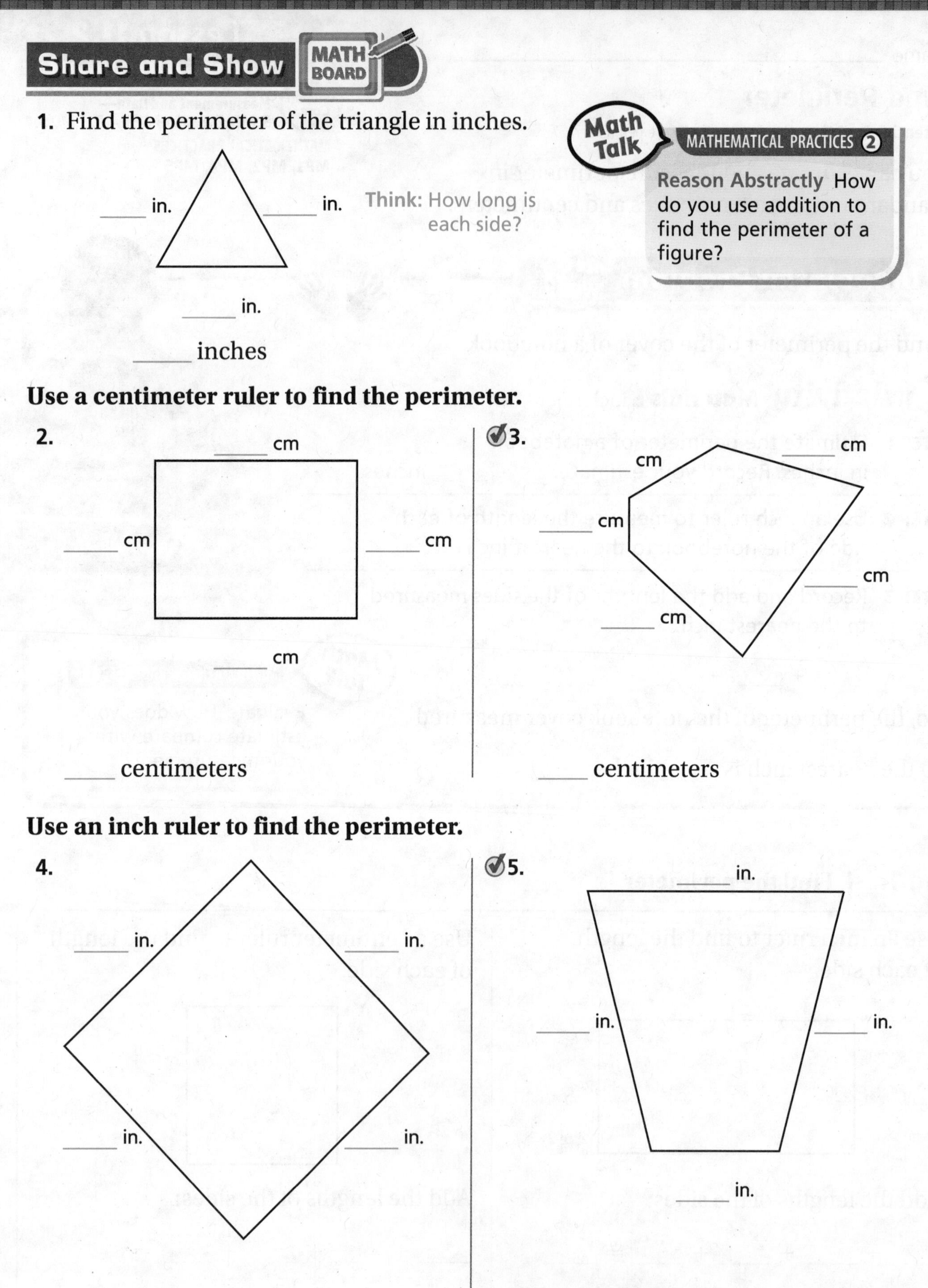

Share and Show MATH BOARD

1. Find the perimeter of the triangle in inches.

____ in. ____ in. ____ in.

Think: How long is each side?

____ inches

Math Talk MATHEMATICAL PRACTICES 2

Reason Abstractly How do you use addition to find the perimeter of a figure?

Use a centimeter ruler to find the perimeter.

2. ____ cm ____ cm ____ cm ____ cm

____ centimeters

3. ____ cm ____ cm ____ cm ____ cm ____ cm

____ centimeters

Use an inch ruler to find the perimeter.

4. ____ in. ____ in. ____ in. ____ in.

____ inches

5. ____ in. ____ in. ____ in. ____ in.

____ inches

Name ____________________

On Your Own

Use a ruler to find the perimeter.

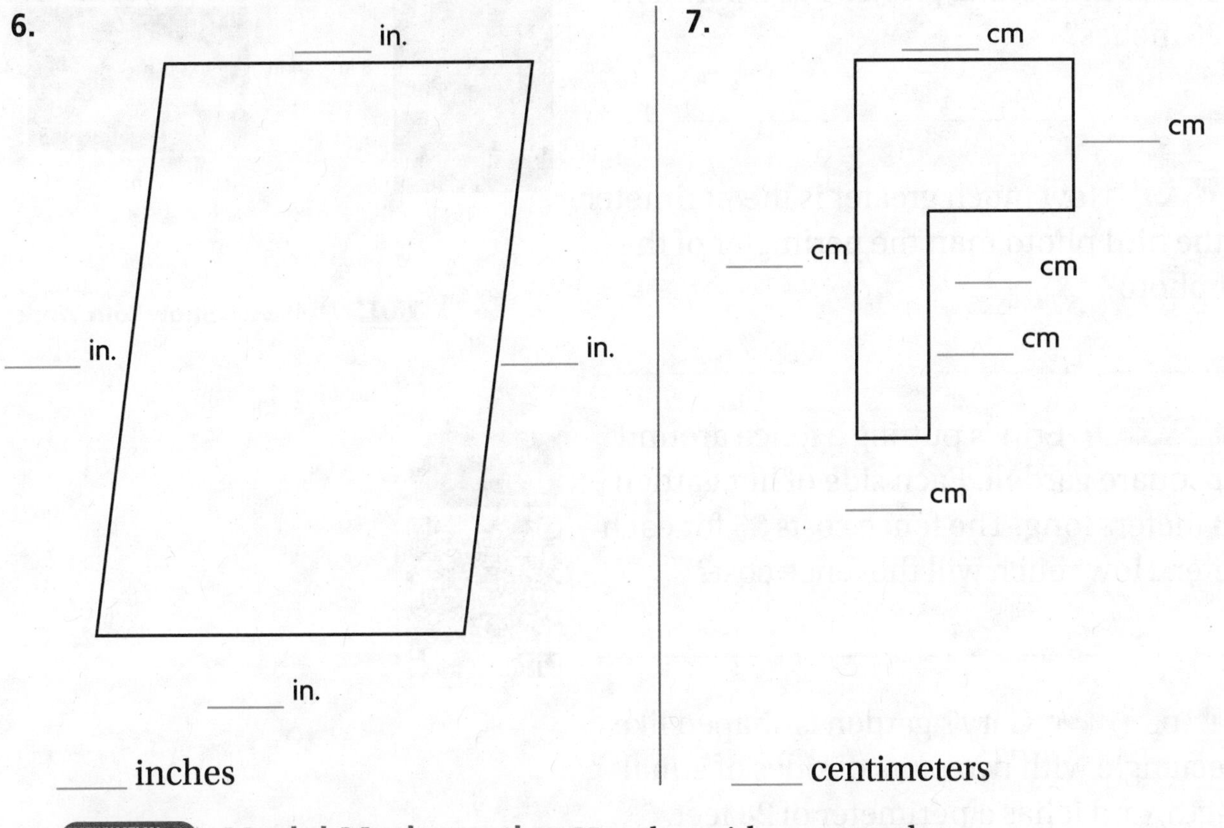

6. ____ inches

7. ____ centimeters

8. MATHEMATICAL PRACTICE 4 **Model Mathematics** Use the grid paper to draw a figure that has a perimeter of 24 centimeters. Label the length of each side.

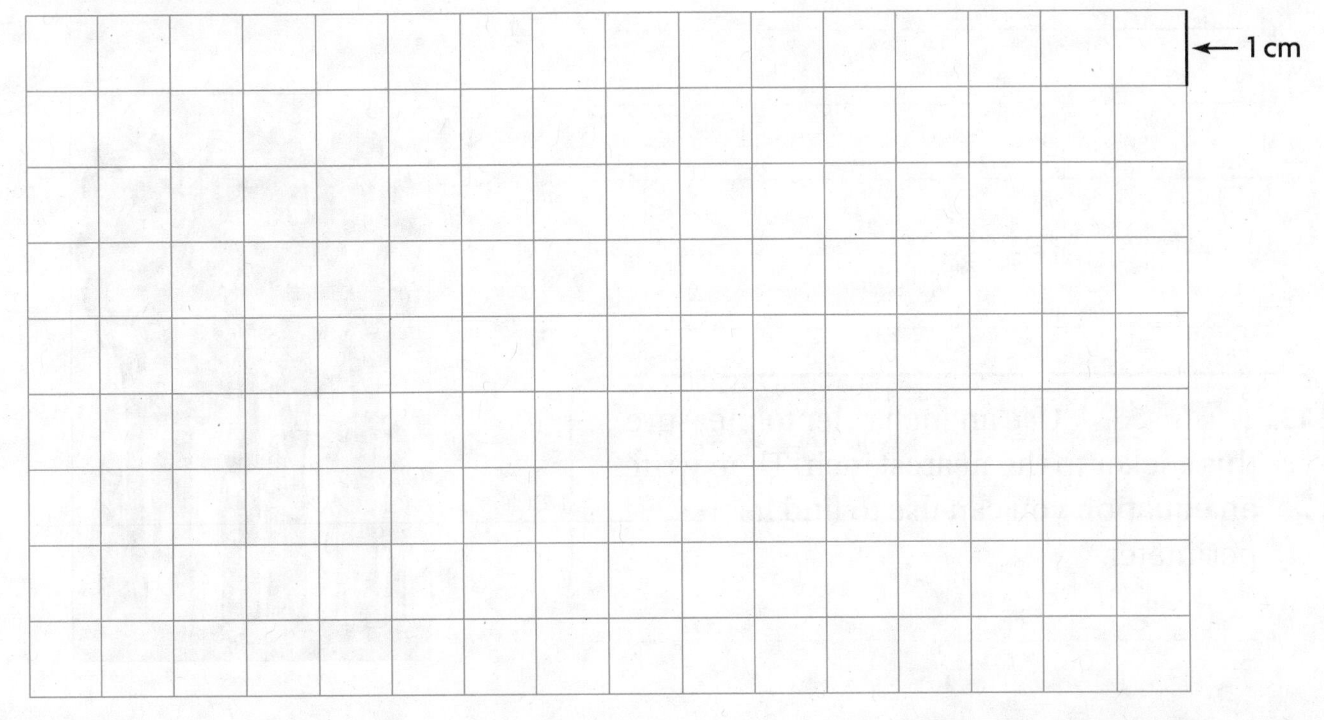

Problem Solving • Applications Real World

Use the photos for 9–10.

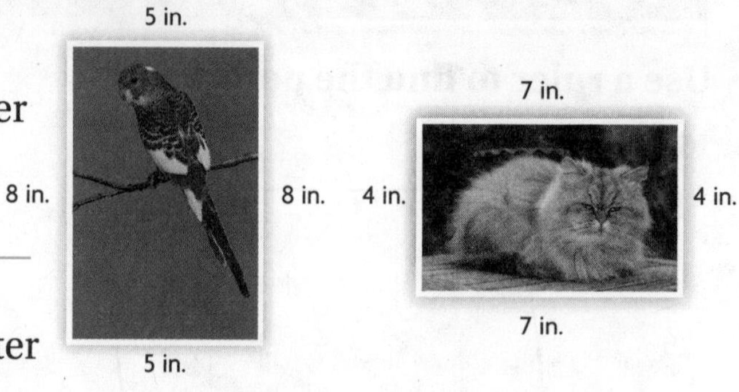

9. Which of the animal photos has a perimeter of 26 inches?

____

10. GO DEEPER How much greater is the perimeter of the bird photo than the perimeter of the cat photo?

____

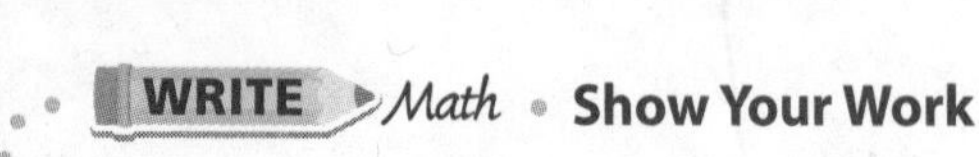

11. THINK SMARTER Erin is putting a fence around her square garden. Each side of her garden is 3 meters long. The fence costs $5 for each meter. How much will the fence cost?

____

Math on the Spot

12. WRITE Math Gary's garden is shaped like a rectangle with two pairs of sides of equal length, and it has a perimeter of 28 feet. Explain how to find the lengths of the other sides if one side measures 10 feet.

____

____

____

____

____

13. THINK SMARTER Use an inch ruler to measure this sticker to the nearest inch. Then write an equation you can use to find its perimeter.

____

Find Perimeter

Common Core COMMON CORE STANDARD—3.MD.D.8
Geometric measurement: recognize perimeter as an attribute of plane figures and distinguish between linear and area measures.

Use a ruler to find the perimeter.

1.

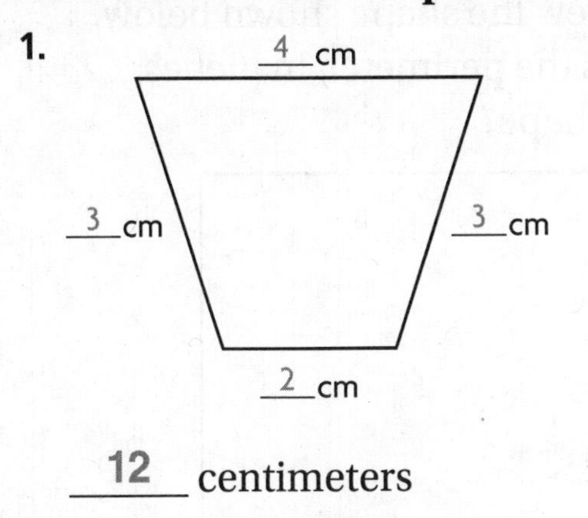

12 centimeters

2.

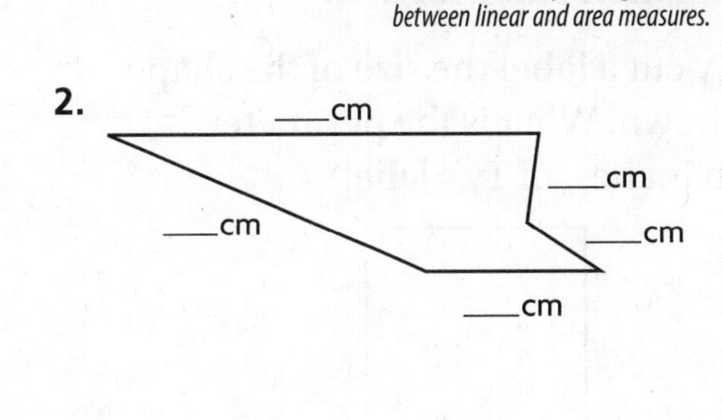

______ centimeters

Draw a picture to solve 3–4.

3. Evan has a square sticker that measures 5 inches on each side. What is the perimeter of the sticker?

____

4. Sophie draws a shape that has 6 sides. Each side is 3 centimeters. What is the perimeter of the shape?

____

5. WRITE *Math* Draw two different figures that each have a perimeter of 20 units.

____

Lesson Check (3.MD.D.8)

Use an inch ruler for 1–2.

1. Ty cut a label the size of the shape shown. What is the perimeter, in inches, of Ty's label?

2. Julie drew the shape shown below. What is the perimeter, in inches, of the shape?

Spiral Review (3.NF.A.3d, 3.MD.A.1, 3.MD.A.2, 3.MD.D.8)

3. What is the perimeter of the shape below?

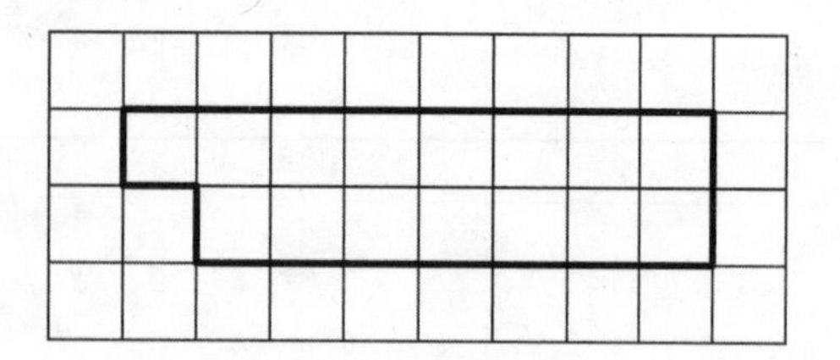

4. Vince arrives for his trumpet lesson after school at the time shown on the clock. What time does Vince arrive for his trumpet lesson?

5. Matthew's small fish tank holds 12 liters. His large fish tank holds 25 liters. How many more liters does his large fish tank hold?

6. Compare. Write <, >, or =.

$\frac{1}{6} \bigcirc \frac{1}{4}$

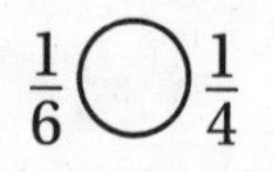

FOR MORE PRACTICE GO TO THE **Personal Math Trainer**

Name ______________________________

ALGEBRA
Lesson 11.3

Algebra • Find Unknown Side Lengths

Essential Question How can you find the unknown length of a side in a plane figure when you know its perimeter?

Common Core **Measurement and Data—3.MD.D.8** *Also 3.NBT.A.2*

MATHEMATICAL PRACTICES
MP3, MP4, MP7

Unlock the Problem Real World

Chen has 27 feet of fencing to put around his garden. He has already used the lengths of fencing shown. How much fencing does he have left for the last side?

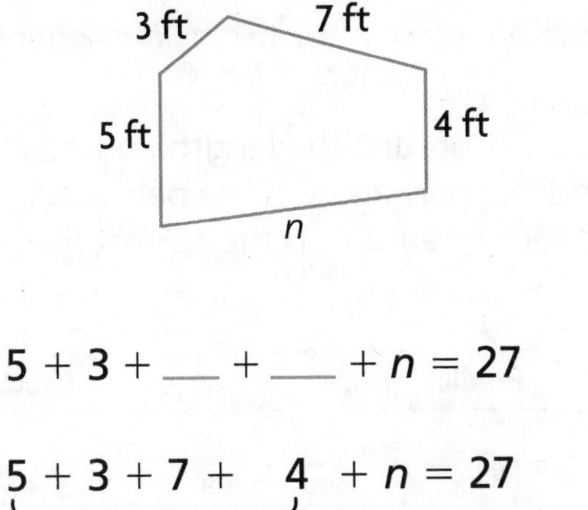

Find the unknown side length.

Write an equation for the perimeter.

$5 + 3 + \_\_ + \_\_ + n = 27$

Think: If I knew the length n, I would add all the side lengths to find the perimeter.

$5 + 3 + 7 + 4 + n = 27$

Add the lengths of the sides you know.

$\_\_ + n = 27$

Think: Addition and subtraction are inverse operations.

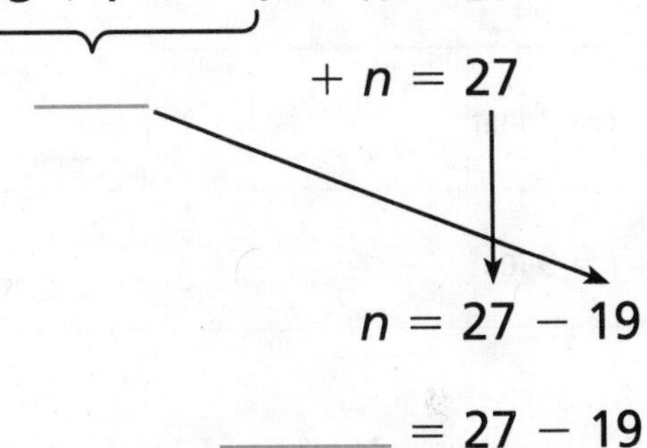

Write a related equation.

$n = 27 - 19$

So, Chen has _______ feet of fencing left.

$\_\_ = 27 - 19$

Math Idea
A symbol or letter can stand for an unknown side length.

Try This!

The perimeter of the figure is 24 meters. Find the unknown side length, w.

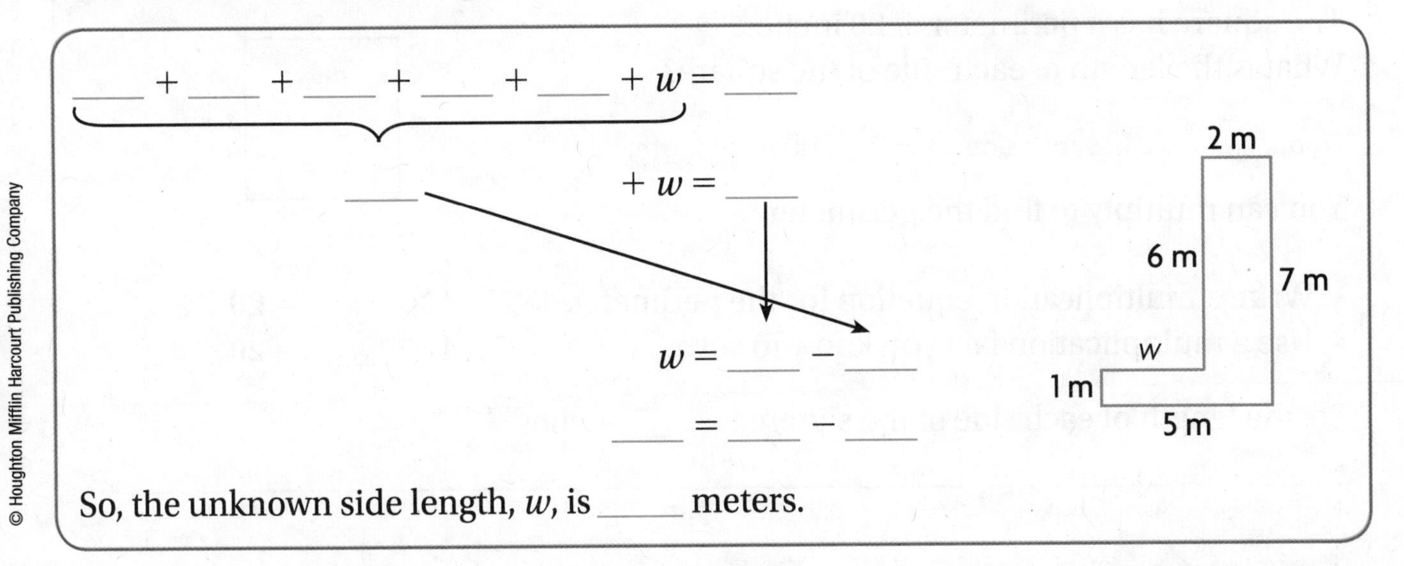

$\_\_ + \_\_ + \_\_ + \_\_ + \_\_ + w = \_\_$

$\_\_ + w = \_\_$

$w = \_\_ - \_\_$

$\_\_ = \_\_ - \_\_$

So, the unknown side length, w, is _____ meters.

Example Find unknown side lengths of a rectangle.

Lauren has a rectangular blanket. The perimeter is 28 feet. The width of the blanket is 5 feet. What is the length of the blanket?

Hint: A rectangle has two pairs of opposite sides that are equal in length.

You can predict the length and add to find the perimeter. If the perimeter is 28 feet, then that is the correct length.

Predict	Check	Does it check?
$l = 7$ feet	5 + ____ + 5 + ____ = ____	**Think:** Perimeter is not 28 feet, so the length does not check.
$l = 8$ feet	5 + ____ + 5 + ____ = ____	**Think:** Perimeter is not 28 feet, so the length does not check.
$l = 9$ feet	5 + ____ + 5 + ____ = ____	**Think:** Perimeter is 28 feet, so the length is correct. ✓

So, the length of the blanket is _____ feet.

Try This! Find unknown side lengths of a square.

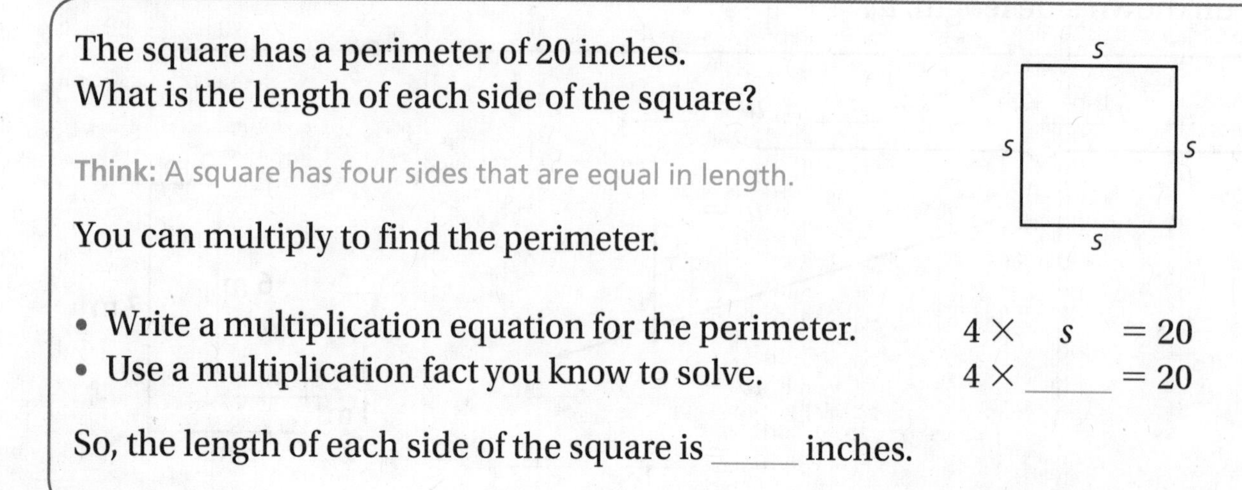

The square has a perimeter of 20 inches.
What is the length of each side of the square?

Think: A square has four sides that are equal in length.

You can multiply to find the perimeter.

- Write a multiplication equation for the perimeter. $4 \times s = 20$
- Use a multiplication fact you know to solve. $4 \times$ ____ $= 20$

So, the length of each side of the square is _____ inches.

Name ________________

Share and Show MATH BOARD

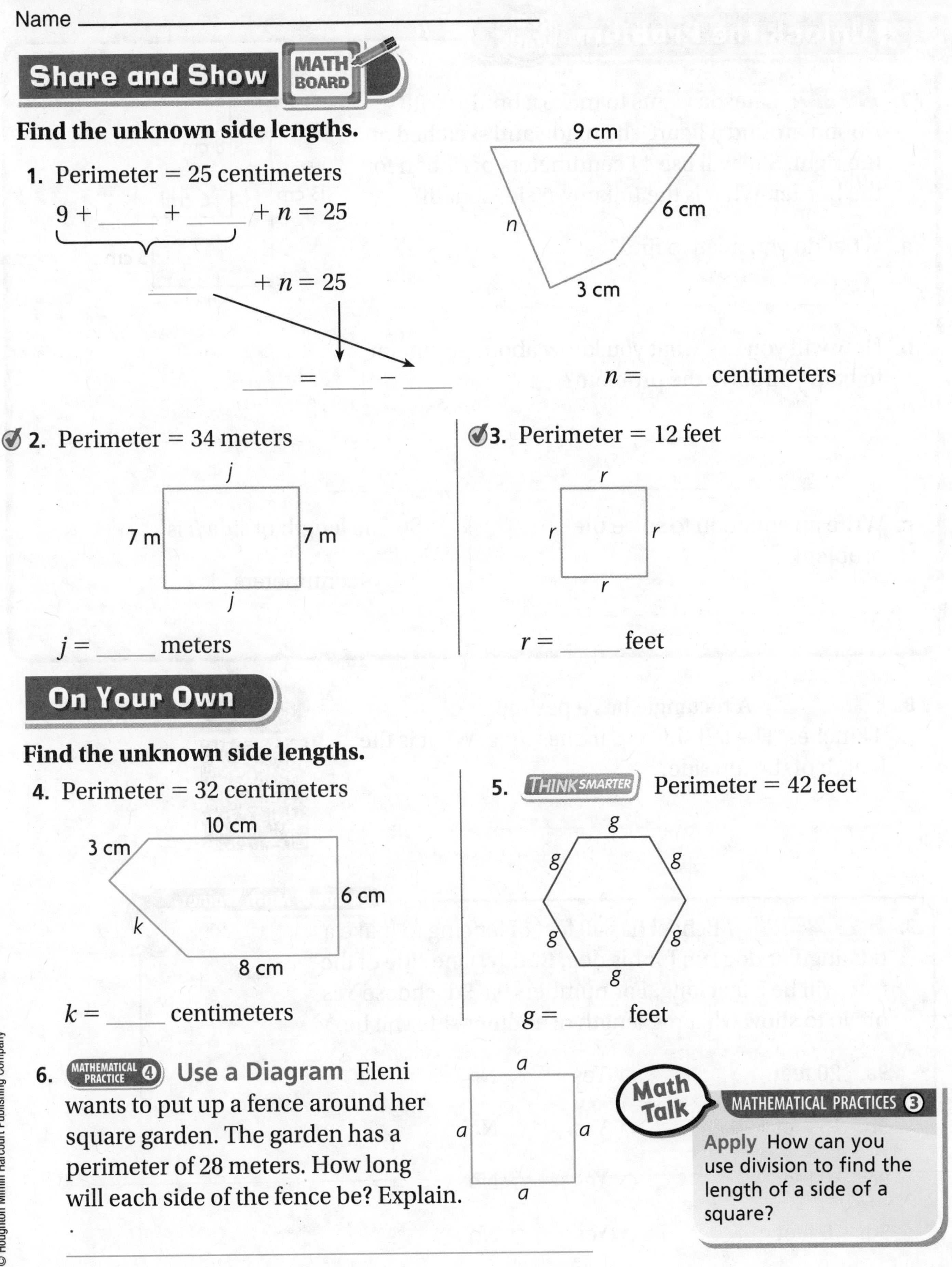

Find the unknown side lengths.

1. Perimeter = 25 centimeters

$9 + \_\_\_ + \_\_\_ + n = 25$

$\_\_\_ + n = 25$

$\_\_\_ = \_\_\_ - \_\_\_$

9 cm, 6 cm, 3 cm, n

$n =$ ____ centimeters

2. Perimeter = 34 meters

j, 7 m, 7 m, j

$j =$ ____ meters

3. Perimeter = 12 feet

r, r, r, r

$r =$ ____ feet

On Your Own

Find the unknown side lengths.

4. Perimeter = 32 centimeters

10 cm, 3 cm, 6 cm, k, 8 cm

$k =$ ____ centimeters

5. THINKSMARTER Perimeter = 42 feet

g, g, g, g, g, g

$g =$ ____ feet

6. MATHEMATICAL PRACTICE 4 **Use a Diagram** Eleni wants to put up a fence around her square garden. The garden has a perimeter of 28 meters. How long will each side of the fence be? Explain.

a, a, a, a

____

____

Math Talk — MATHEMATICAL PRACTICES 3

Apply How can you use division to find the length of a side of a square?

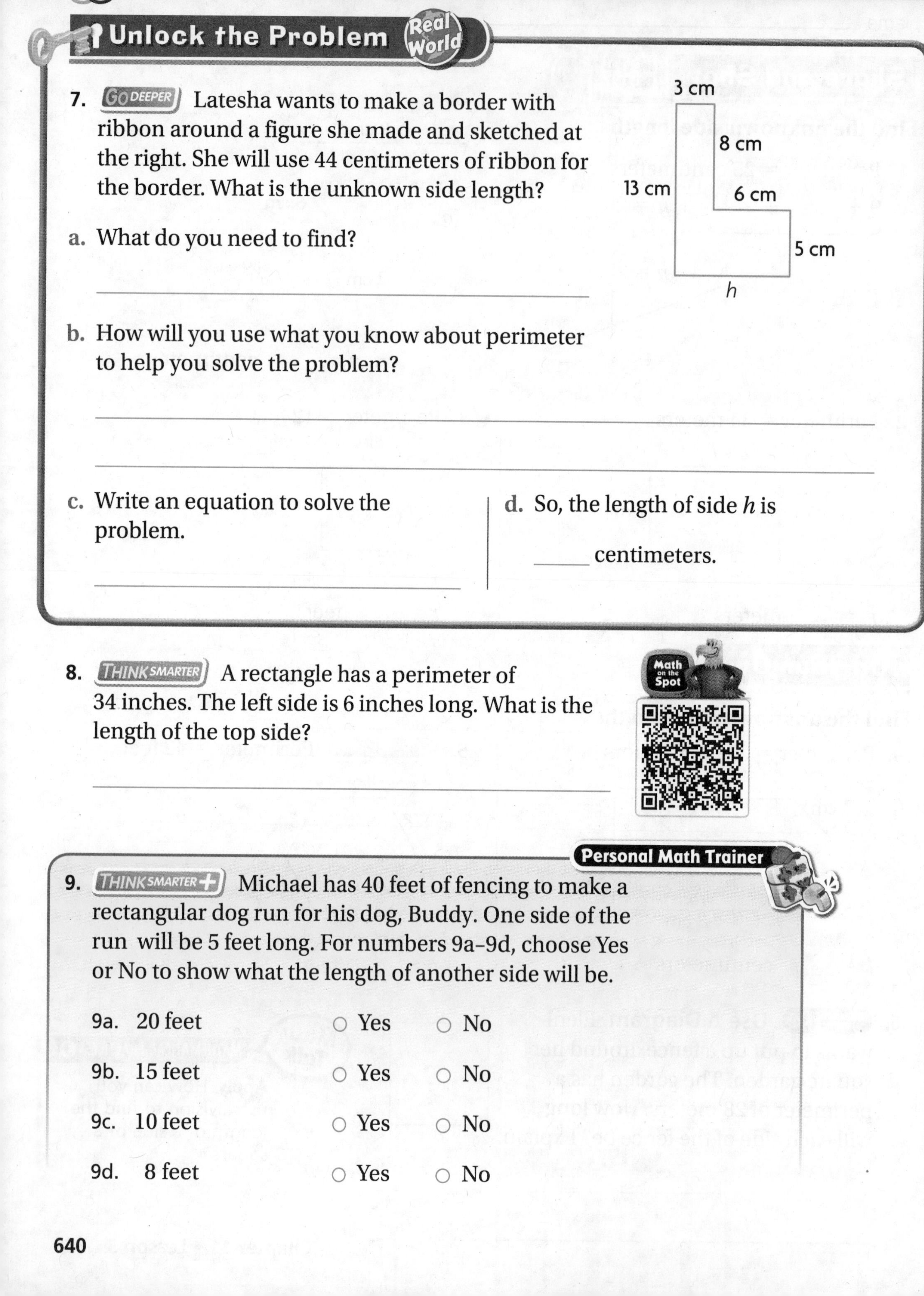

Unlock the Problem — Real World

7. GO DEEPER Latesha wants to make a border with ribbon around a figure she made and sketched at the right. She will use 44 centimeters of ribbon for the border. What is the unknown side length?

a. What do you need to find?

____

b. How will you use what you know about perimeter to help you solve the problem?

____

____

c. Write an equation to solve the problem.

____

d. So, the length of side h is ______ centimeters.

8. THINK SMARTER A rectangle has a perimeter of 34 inches. The left side is 6 inches long. What is the length of the top side?

____

9. THINK SMARTER+ Michael has 40 feet of fencing to make a rectangular dog run for his dog, Buddy. One side of the run will be 5 feet long. For numbers 9a–9d, choose Yes or No to show what the length of another side will be.

9a.	20 feet	○ Yes	○ No
9b.	15 feet	○ Yes	○ No
9c.	10 feet	○ Yes	○ No
9d.	8 feet	○ Yes	○ No

Name ______________________

Find Unknown Side Lengths

COMMON CORE STANDARD—3.MD.D.8
Geometric measurement: recognize perimeter as an attribute of plane figures and distinguish between linear and area measures.

Find the unknown side lengths.

1. Perimeter = 33 centimeters

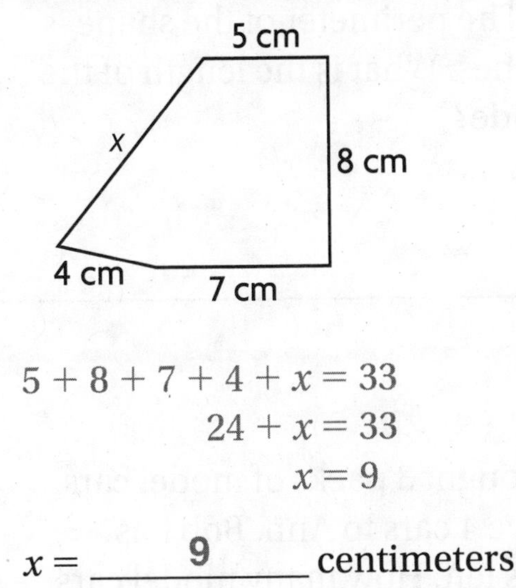

$5 + 8 + 7 + 4 + x = 33$

$24 + x = 33$

$x = 9$

$x =$ ___9___ centimeters

2. Perimeter = 92 inches

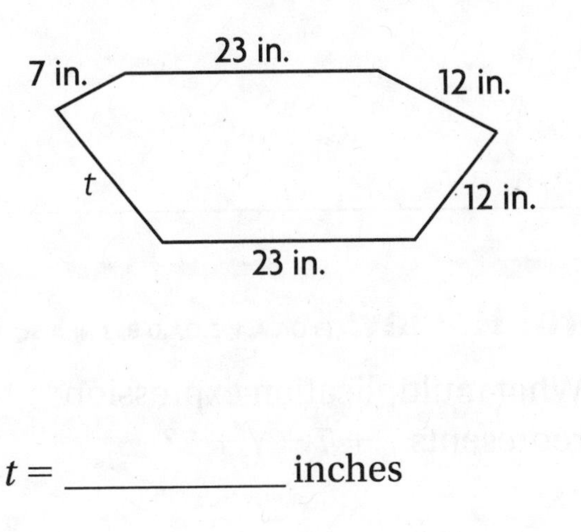

$t =$ __________ inches

Problem Solving Real World

3. Steven has a rectangular rug with a perimeter of 16 feet. The width of the rug is 5 feet. What is the length of the rug?

____

4. Kerstin has a square tile. The perimeter of the tile is 32 inches. What is the length of each side of the tile?

____

5. WRITE Math Explain how to write and solve an equation to find an unknown side length of a rectangle when given the perimeter.

____

____

____

Lesson Check (3.MD.D.8)

1. Jesse is putting a ribbon around a square frame. He uses 24 inches of ribbon. How long is each side of the frame?

2. Davia draws a shape with 5 sides. Two sides are each 5 inches long. Two other sides are each 4 inches long. The perimeter of the shape is 27 inches. What is the length of the fifth side?

Spiral Review (3.OA.A.1, 3.OA.D.8, 3.NF.A.3c, 3.MD.A.1)

3. What multiplication expression represents $7 + 7 + 7 + 7$?

4. Bob bought 3 packs of model cars. He gave 4 cars to Ann. Bob has 11 cars left. How many model cars were in each pack?

5. Randy read a book in the afternoon. He looked at his watch when he started and finished reading. How long did Randy read?

6. What fraction and whole number does the model represent?

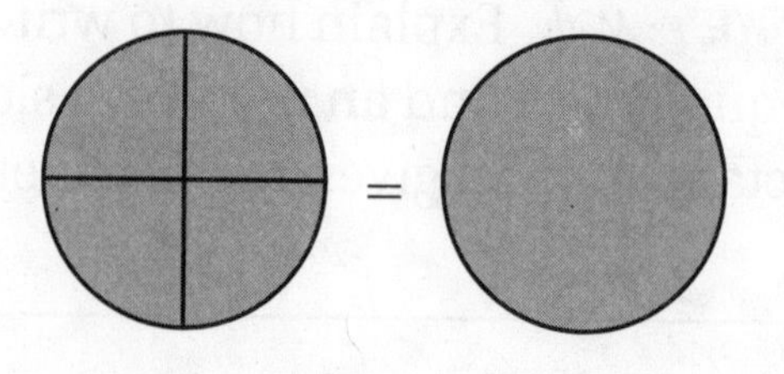

_______ = _______

FOR MORE PRACTICE GO TO THE Personal Math Trainer

Name ____________________

Understand Area

Essential Question How is finding the area of a figure different from finding the perimeter of a figure?

Common Core **Measurement and Data—3.MD.C.5, 3.MD.C.5a** *Also 3.MD.C.5b, 3.MD.C.6, 3.MD.D.8*

MATHEMATICAL PRACTICES MP2, MP3, MP5, MP8

Unlock the Problem — Real World

CONNECT You learned that perimeter is the distance around a figure. It is measured in linear units, or units that are used to measure the distance between two points.

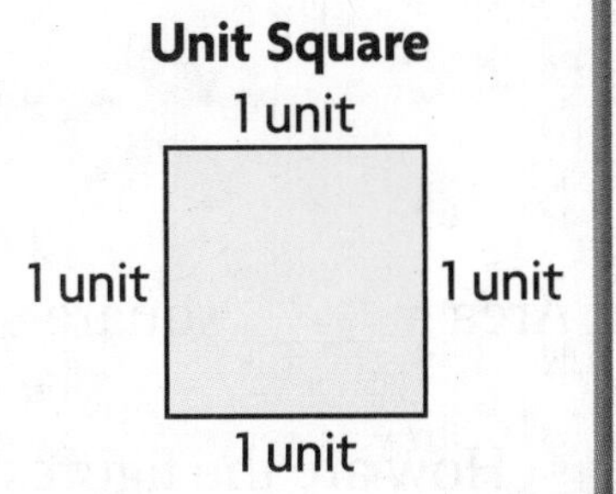

Area is the measure of the number of unit squares needed to cover a flat surface. A **unit square** is a square with a side length of 1 unit. It has an area of 1 **square unit**.

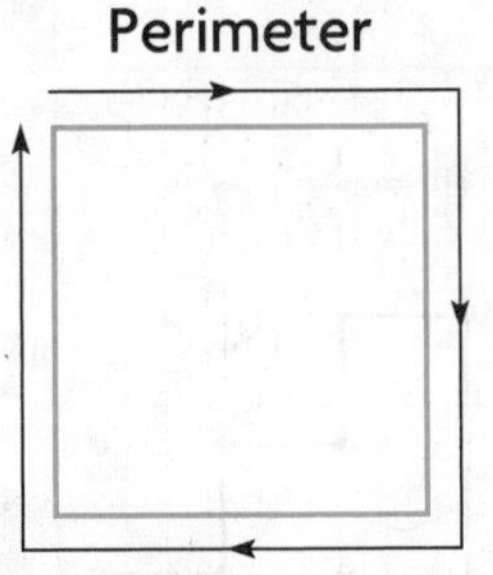

1 unit + 1 unit + 1 unit + 1 unit = 4 units

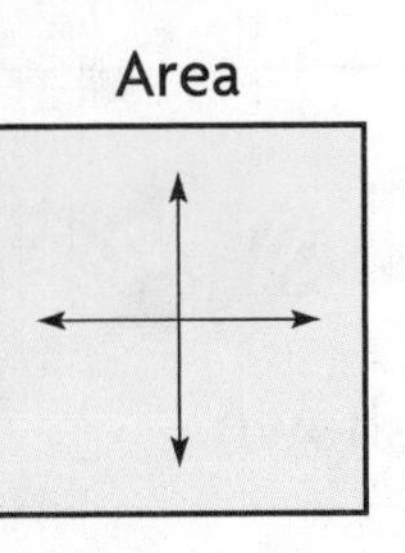

1 square unit

Math Idea

You can count the number of units on each side of a figure to find its perimeter. You can count the number of unit squares inside a figure to find its area in square units.

Activity **Materials** ■ geoboard ■ rubber bands

A Use your geoboard to form a figure made from 2 unit squares. Record the figure on this dot paper.

What is the area of this figure?

Area = _____ square units

B Change the rubber band so that the figure is made from 3 unit squares. Record the figure on this dot paper.

What is the area of this figure?

Area = _____ square units

MATHEMATICAL PRACTICES 3

Compare Representations For B, did your figure look like your classmate's figure?

Try This! **Draw three different figures that are each made from 4 unit squares. Find the area of each figure.**

Figure 1

Area = _____ square units

Figure 2

Area = _____ square units

Figure 3

Area = _____ square units

- How are the figures the same? How are the figures different?

____

Share and Show

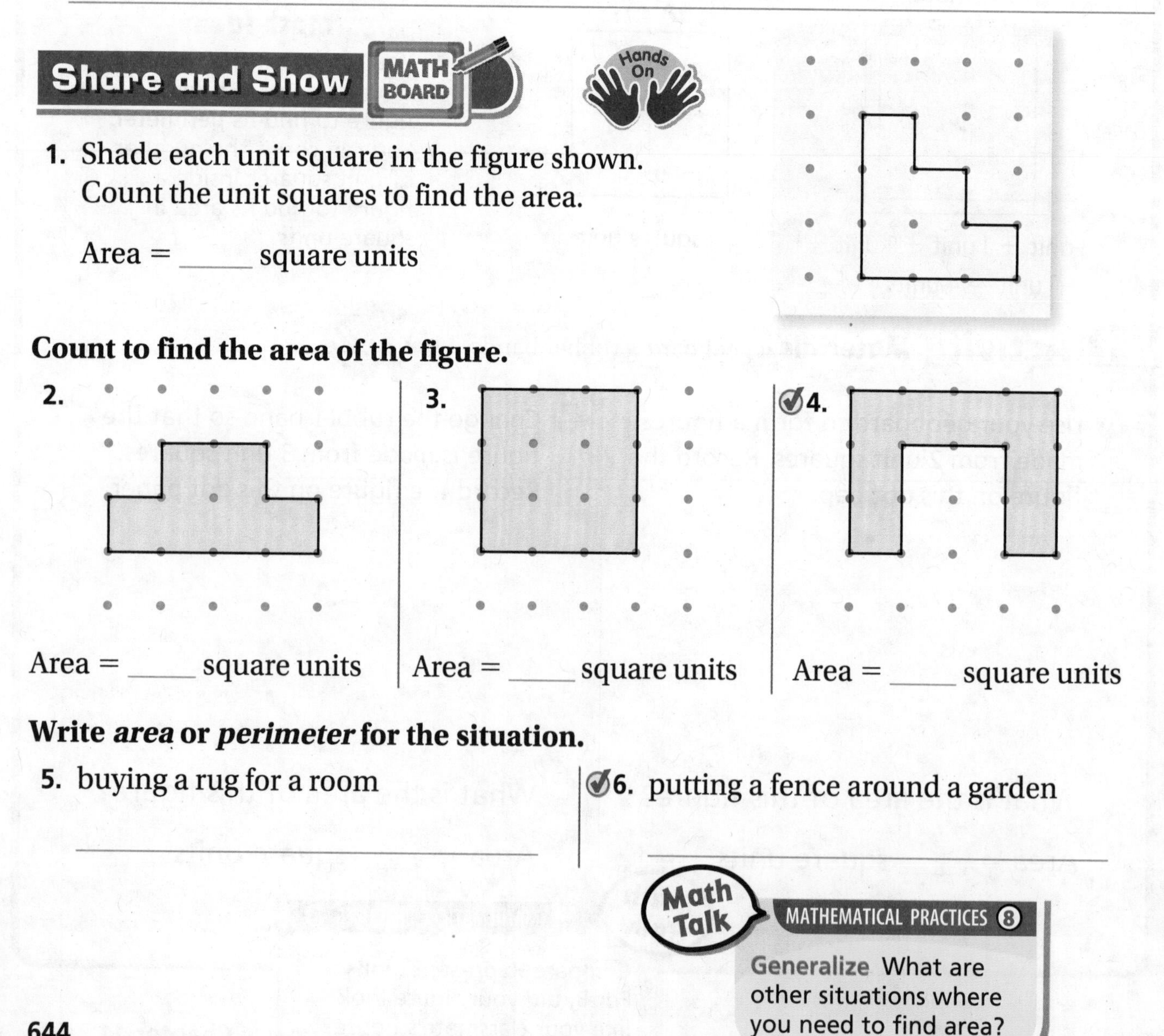

1. Shade each unit square in the figure shown. Count the unit squares to find the area.

Area = _____ square units

Count to find the area of the figure.

2. Area = _____ square units

3. Area = _____ square units

✓4. Area = _____ square units

Write *area* or *perimeter* for the situation.

5. buying a rug for a room

____

✓6. putting a fence around a garden

____

Math Talk

MATHEMATICAL PRACTICES 8

Generalize What are other situations where you need to find area?

Name ____

On Your Own

Count to find the area of the figure.

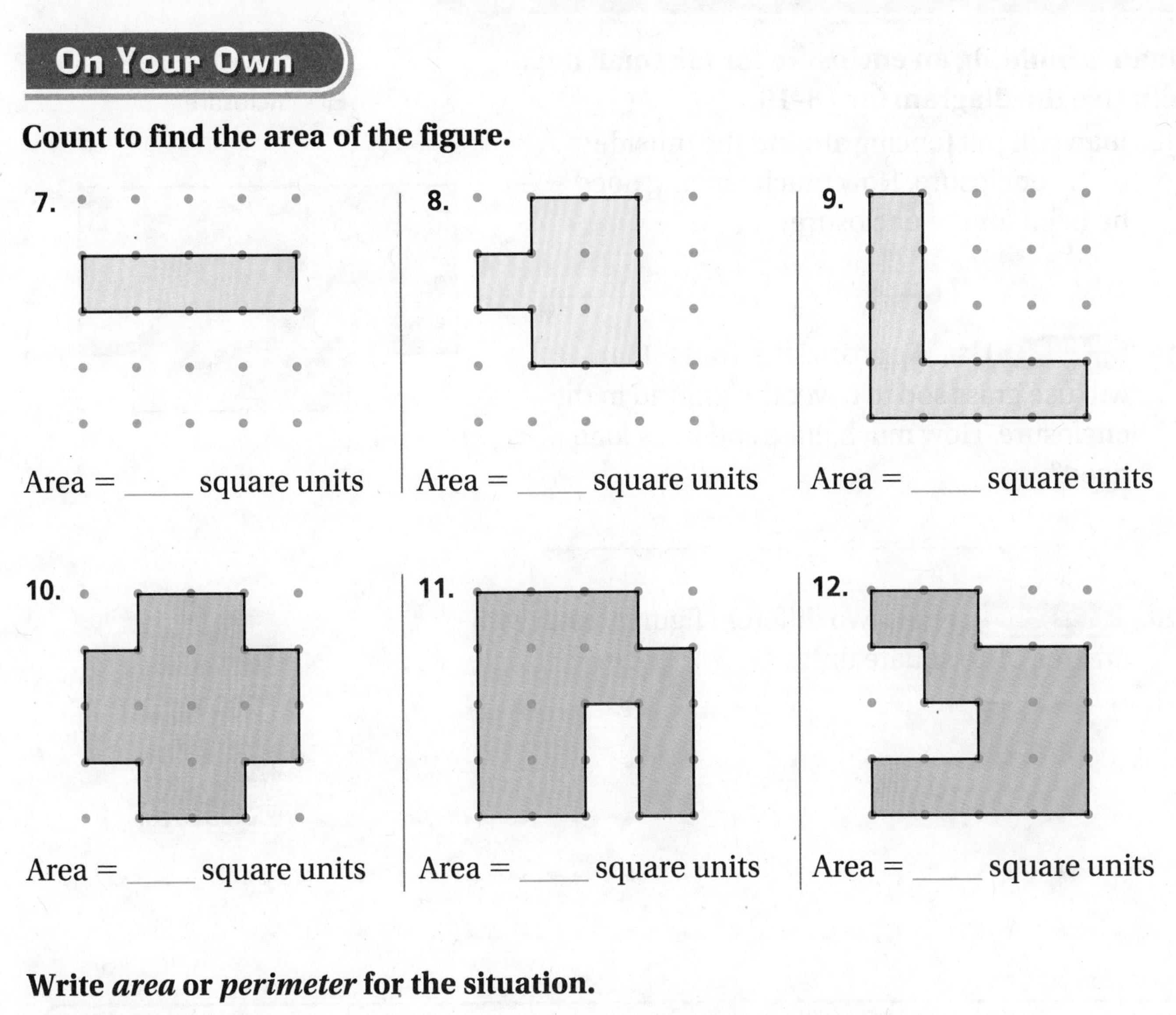

Write *area* or *perimeter* for the situation.

13. painting a wall

____

14. covering a patio with tiles

____

15. putting a wallpaper border around a room

____

16. gluing a ribbon around a picture frame

____

17. GO DEEPER Nicole's mother put tiles on a section of their kitchen floor. The section included 5 rows with 4 tiles in each row. Each tile cost $2. How much money did Nicole's mother spend on the tiles?

____

Problem Solving • Applications

Juan is building an enclosure for his small dog, Eli. Use the diagram for 18–19.

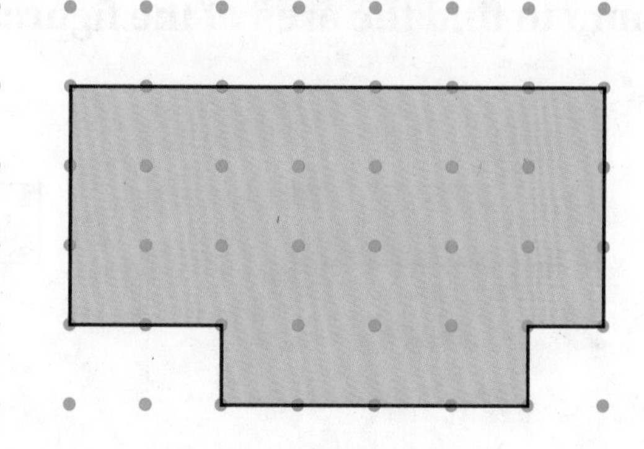

18. Juan will put fencing around the outside of the enclosure. How much fencing does he need for the enclosure?

____

19. MATHEMATICAL PRACTICE 5 **Use Appropriate Tools** Juan will use grass sod to cover the ground in the enclosure. How much grass sod does Juan need?

____

20. THINKSMARTER Draw two different figures, each with an area of 10 square units.

21. THINKSMARTER What is the perimeter and area of this figure? Explain how you found the answer.

Perimeter ______ units

Area ______ square units

____

____

____

____

Name ______________________

Understand Area

Common Core

COMMON CORE STANDARDS—3.MD.C.5, 3.MD.C.5a *Geometric measurement: understand concepts of area and relate area to multiplication and to addition.*

Count to find the area for the shape.

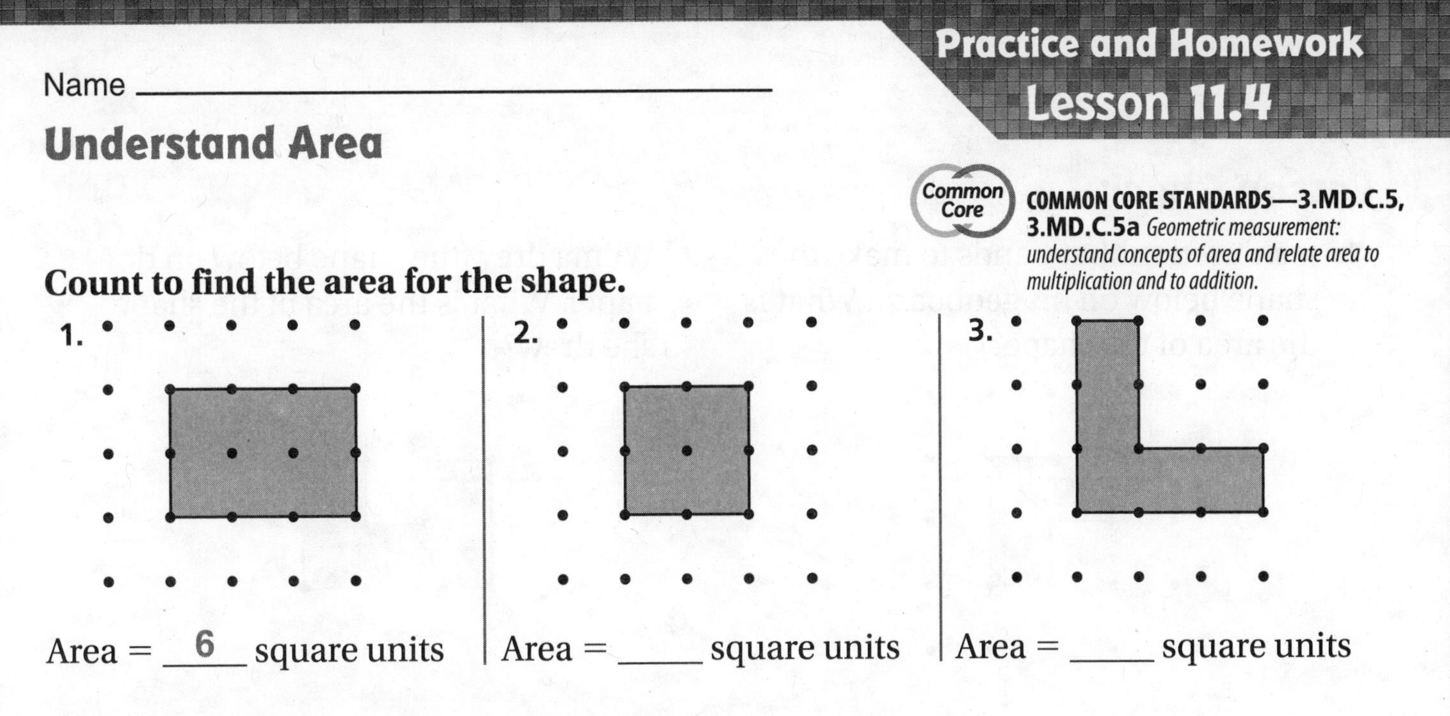

1. Area = __6__ square units
2. Area = _____ square units
3. Area = _____ square units

Write *area* or *perimeter* for each situation.

4. carpeting a floor

____

5. fencing a garden

____

Problem Solving Real World

Use the diagram for 6–7.

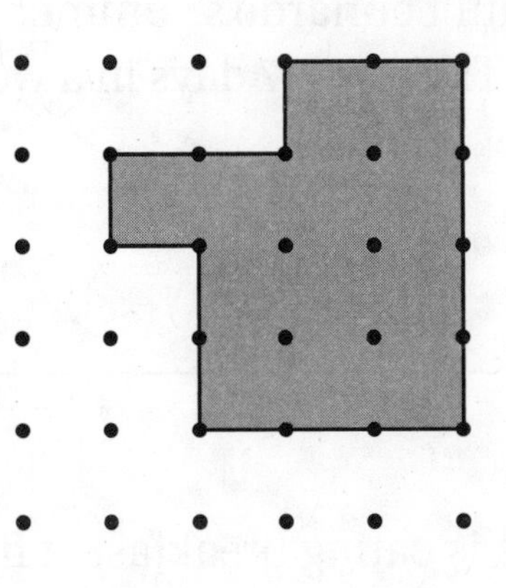

6. Roberto is building a platform for his model railroad. What is the area of the platform?

____

7. Roberto will put a border around the edges of the platform. How much border will he need?

____

8. WRITE Math Draw a rectangle using dot paper. Find the area, and explain how you found your answer.

____

Lesson Check (3.MD.C.5, 3.MD.C.5a)

1. Josh used rubber bands to make the shape below on his geoboard. What is the area of the shape?

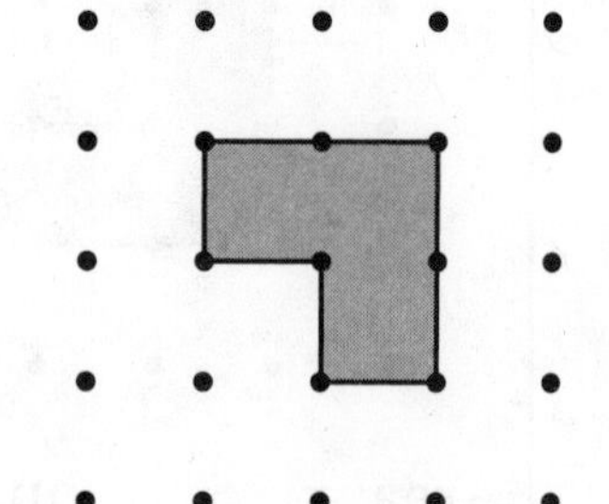

2. Wilma drew the shape below on dot paper. What is the area of the shape she drew?

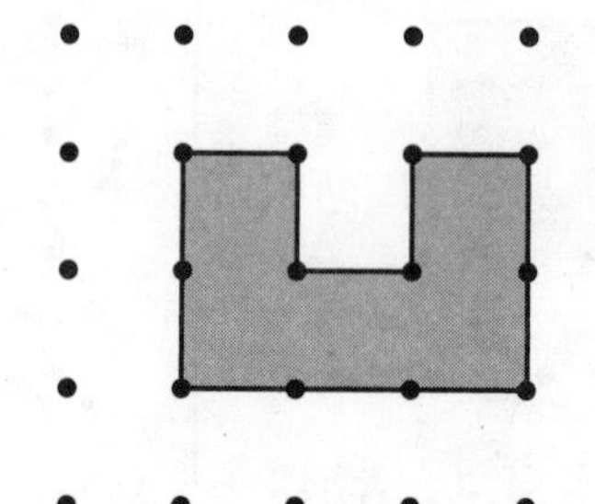

Spiral Review (3.OA.C.7, 3.NF.A.1, 3.MD.A.1, 3.MD.A.2)

3. Leonardo knows it is 42 days until summer break. How many weeks is it until Leonardo's summer break? (Hint: There are 7 days in a week.)

4. Nan cut a submarine sandwich into 4 equal parts and ate one part. What fraction represents the part of the sandwich Nan ate?

5. Wanda is eating breakfast at fifteen minutes before eight. What time is this? Use A.M. or P.M.

6. Dick has 2 bags of dog food. Each bag contains 5 kilograms of food. How many kilograms of food does Dick have in all?

FOR MORE PRACTICE GO TO THE Personal Math Trainer

Name ______

Measure Area

Essential Question How can you find the area of a plane figure?

Common Core **Measurement and Data—3.MD.C.5b, 3.MD.C.6** *Also 3.MD.C.5, 3.MD.C.5a, 3.MD.C.7, 3.MD.C.7a*

MATHEMATICAL PRACTICES
MP2, MP4, MP7

Unlock the Problem

Hands On

Jaime is measuring the area of the rectangles with 1-inch square tiles.

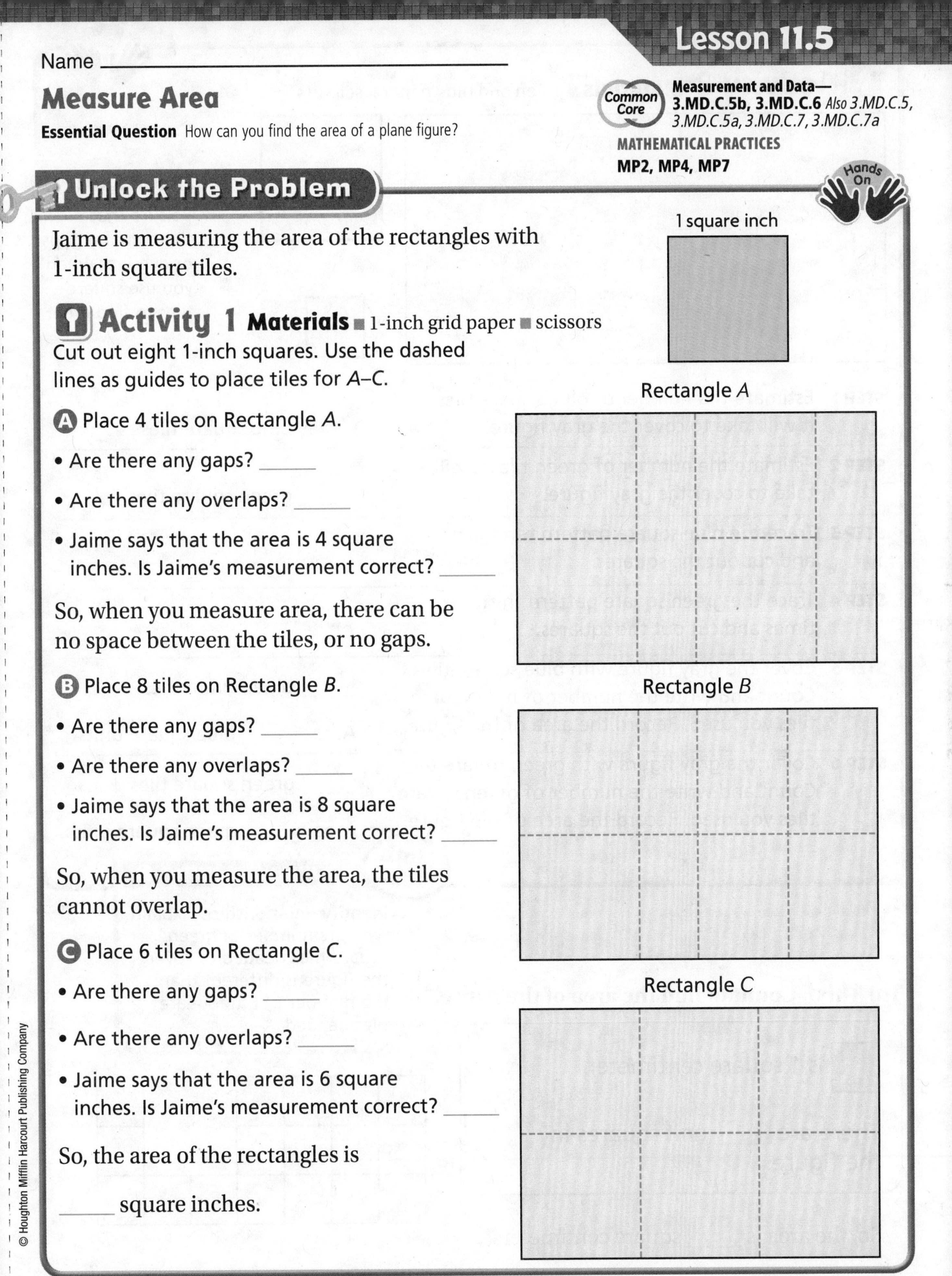

Activity 1 **Materials** ■ 1-inch grid paper ■ scissors

Cut out eight 1-inch squares. Use the dashed lines as guides to place tiles for *A–C*.

A Place 4 tiles on Rectangle *A*.

- Are there any gaps? ______
- Are there any overlaps? ______
- Jaime says that the area is 4 square inches. Is Jaime's measurement correct? ______

So, when you measure area, there can be no space between the tiles, or no gaps.

B Place 8 tiles on Rectangle *B*.

- Are there any gaps? ______
- Are there any overlaps? ______
- Jaime says that the area is 8 square inches. Is Jaime's measurement correct? ______

So, when you measure the area, the tiles cannot overlap.

C Place 6 tiles on Rectangle C.

- Are there any gaps? ______
- Are there any overlaps? ______
- Jaime says that the area is 6 square inches. Is Jaime's measurement correct? ______

So, the area of the rectangles is

______ square inches.

Hands On

Activity 2 **Materials** ■ green and blue paper ■ scissors

ERROR Alert

Be sure that there are no gaps or overlaps when you use square tiles to find area.

STEP 1 Estimate the number of blue square tiles it will take to cover the gray figure. ______ blue square tiles

STEP 2 Estimate the number of green tiles it will take to cover the gray figure. ______ green square tiles

STEP 3 Trace the blue square pattern ten times and cut out the squares.

STEP 4 Trace the green square pattern thirty-six times and cut out the squares.

STEP 5 Cover the gray figure with blue square tiles. Count and write the number of blue square tiles you used. Record the area of the figure.

______ blue square tiles

Area = ____ blue square units

STEP 6 Cover the gray figure with green square tiles. Count and write the number of green square tiles you used. Record the area of the figure.

______ green square tiles

Area = ____ green square units

Math Talk MATHEMATICAL PRACTICES 7

Identify Relationships Explain why the number of green square tiles needed to cover the figure is different than the number of blue square tiles needed.

Try This! Count to find the area of the figure.

▢ is 1 square centimeter.

There are _____ unit squares in the figure.

1
2

So, the area is _____ square centimeters.

Name ________________

Share and Show MATH BOARD

1. Count to find the area of the figure. Each unit square is 1 square centimeter.

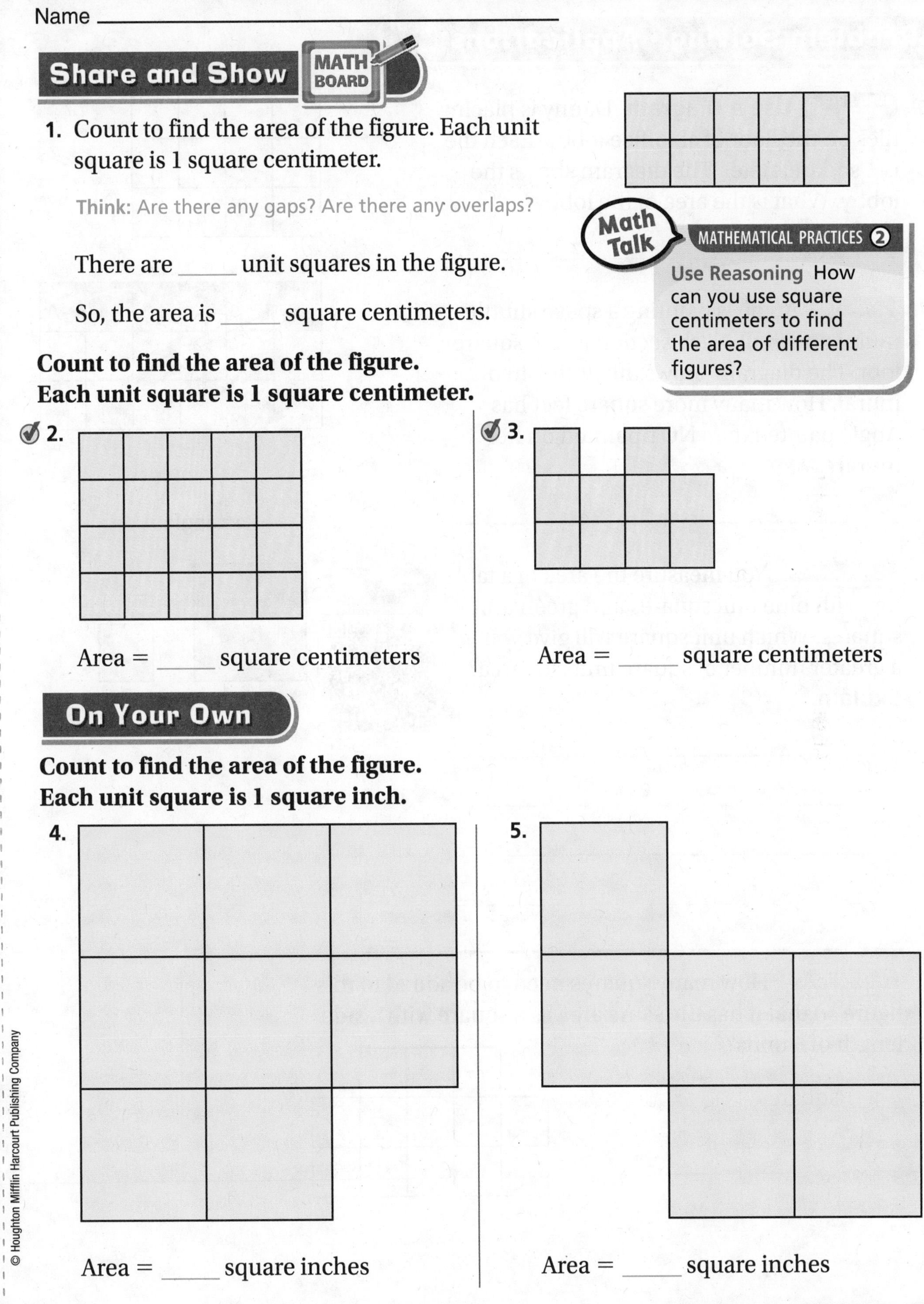

Think: Are there any gaps? Are there any overlaps?

There are _____ unit squares in the figure.

So, the area is _____ square centimeters.

Math Talk — MATHEMATICAL PRACTICES 2

Use Reasoning How can you use square centimeters to find the area of different figures?

Count to find the area of the figure. Each unit square is 1 square centimeter.

2. Area = _____ square centimeters

3. Area = _____ square centimeters

On Your Own

Count to find the area of the figure. Each unit square is 1 square inch.

4. Area = _____ square inches

5. Area = _____ square inches

Problem Solving • Applications Real World

6. MATHEMATICAL PRACTICE 4 **Use a Diagram** Danny is placing tiles on the floor of an office lobby. Each tile is 1 square meter. The diagram shows the lobby. What is the area of the lobby?

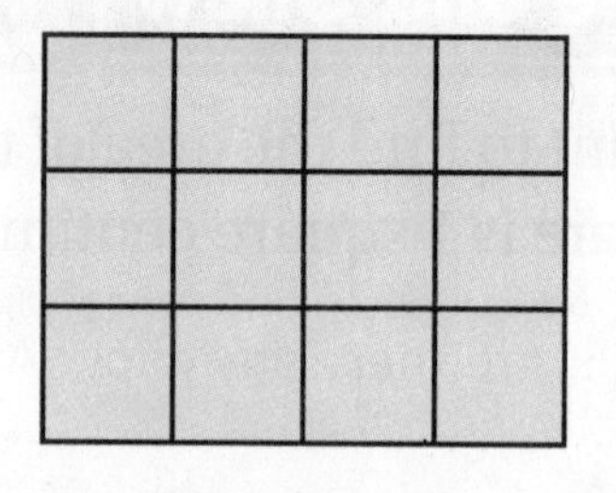

____

7. GO DEEPER Angie is painting a space shuttle mural on a wall. Each section is one square foot. The diagram shows the unfinished mural. How many more square feet has Angie painted than NOT painted on her mural?

____

8. THINK SMARTER You measure the area of a table top with blue unit squares and green unit squares. Which unit square will give you a greater number of square units for area? **Explain.**

Math on the Spot

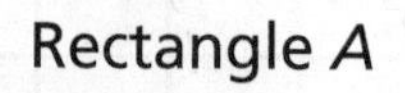

Rectangle *A*

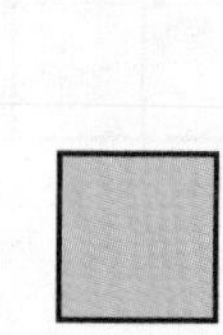

____

____

____

____

9. THINK SMARTER How many squares need to be added to this figure so that it has the same area as a square with a side length of 5 units?

_____ squares

Name ____________________

Measure Area

COMMON CORE STANDARDS—3.MD.C.5b, 3.MD.C.6 *Geometric measurement: understand concepts of area and relate area to multiplication and to addition.*

Count to find the area of the shape. Each unit square is 1 square centimeter.

1. 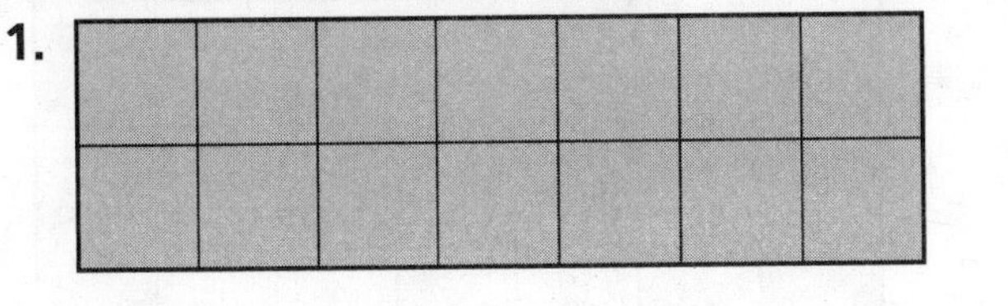

Area = 14 square centimeters

2. 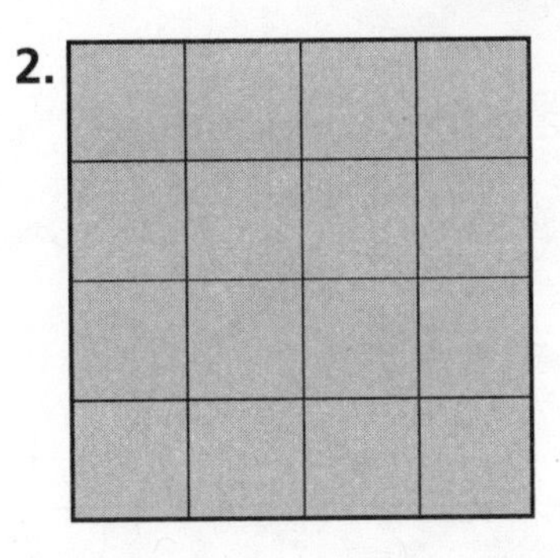

Area = ______ square centimeters

3. 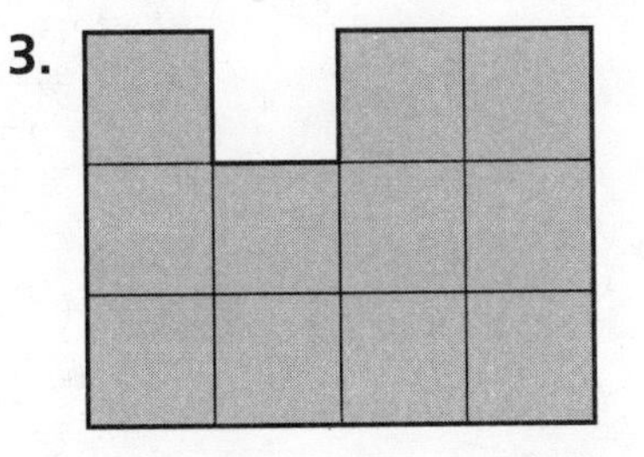

Area = ______ square centimeters

4. 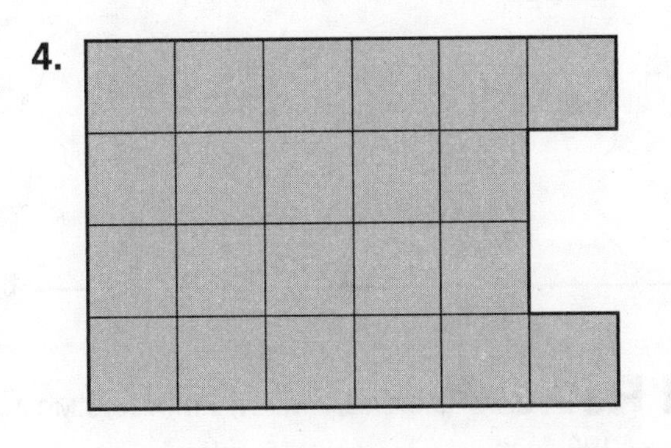

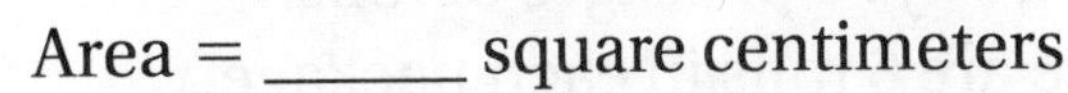

Problem Solving Real World

Alan is painting his deck gray. Use the diagram at the right for 5. Each unit square is 1 square meter.

Alan's Deck

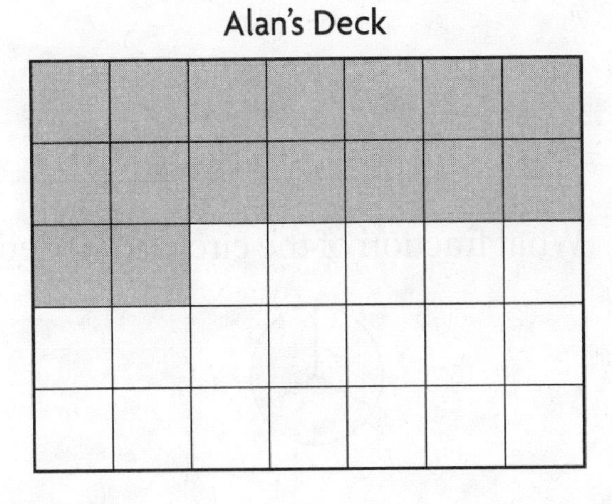

5. What is the area of the deck that Alan has already painted gray?

____

6. WRITE *Math* Explain how to find the area of a figure using square tiles.

____

____

Lesson Check (3.MD.C.5b, 3.MD.C.6)

Each unit square in the diagram is 1 square foot.

1. How many square feet are shaded?

____

2. What is the area that has NOT been shaded?

____

Spiral Review (3.OA.A.3, 3.NF.A.1, 3.NF.A.3b, 3.MD.A.2)

3. Sonya buys 6 packages of rolls. There are 6 rolls in each package. How many rolls does Sonya buy?

____

4. Charlie mixed 6 liters of juice with 2 liters of soda to make fruit punch. How many liters of fruit punch did Charlie make?

____

5. What fraction of the circle is shaded?

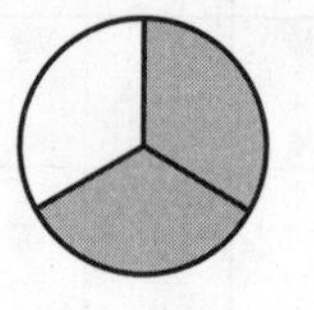

____

6. Use the model on the right to name a fraction that is equivalent to $\frac{1}{2}$.

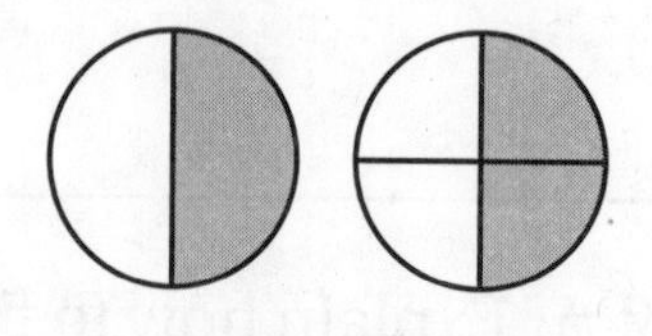

____

FOR MORE PRACTICE GO TO THE **Personal Math Trainer**

Name ______________________

Use Area Models

Essential Question Why can you multiply to find the area of a rectangle?

Common Core **Measurement and Data—3.MD.C.7, 3.MD.C.7a** *Also 3.MD.C.5, 3.MD.C.5a, 3.MD.C.5b, 3.MD.C.6, 3.MD.C.7b, 3.OA.A.3, 3.OA.C.7, 3.NBT.A.2*

MATHEMATICAL PRACTICES
MP1, MP4, MP5, MP6

Unlock the Problem — Real World

Cristina has a garden that is shaped like the rectangle below. Each unit square represents 1 square meter. What is the area of her garden?

- Circle the shape of the garden.

One Way Count unit squares.

Count the number of unit squares in all.

There are _____ unit squares.

So, the area is _____ square meters.

Other Ways

A Use repeated addition.

Count the number of rows. Count the number of unit squares in each row.

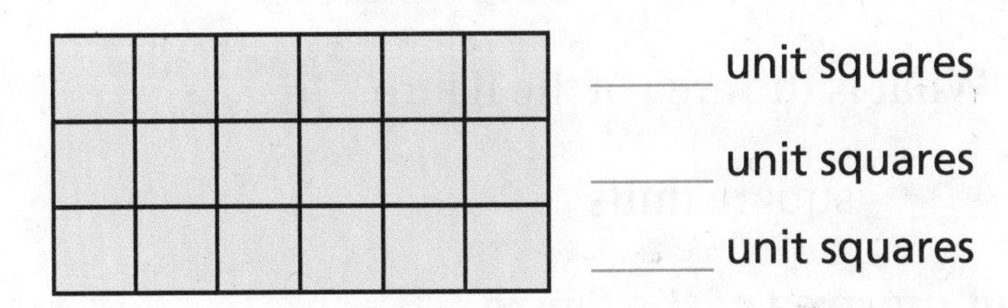

_____ unit squares
_____ unit squares
_____ unit squares

_____ rows of _____ = ■

Write an addition equation.

_____ + _____ + _____ = _____

So, the area is _____ square meters.

B Use multiplication.

Count the number of rows. Count the number of unit squares in each row.

_____ rows of _____ = ■

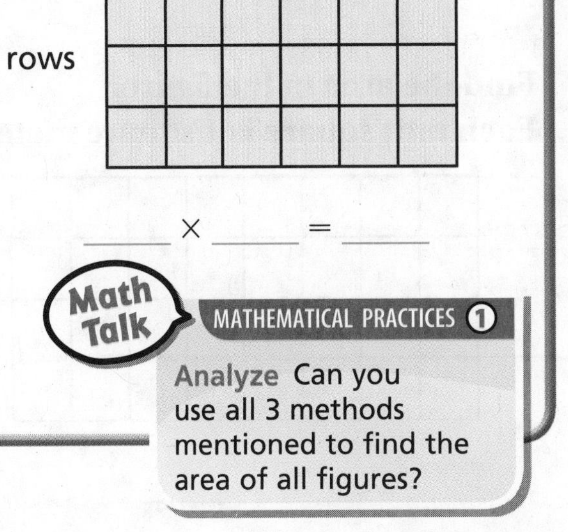

_____ unit squares in each row

_____ rows

_____ × _____ = _____

This rectangle is like an array. How do you find the total number of squares in an array?

____

Write a multiplication equation.

So, the area is _____ square meters.

Math Talk MATHEMATICAL PRACTICES 1

Analyze Can you use all 3 methods mentioned to find the area of all figures?

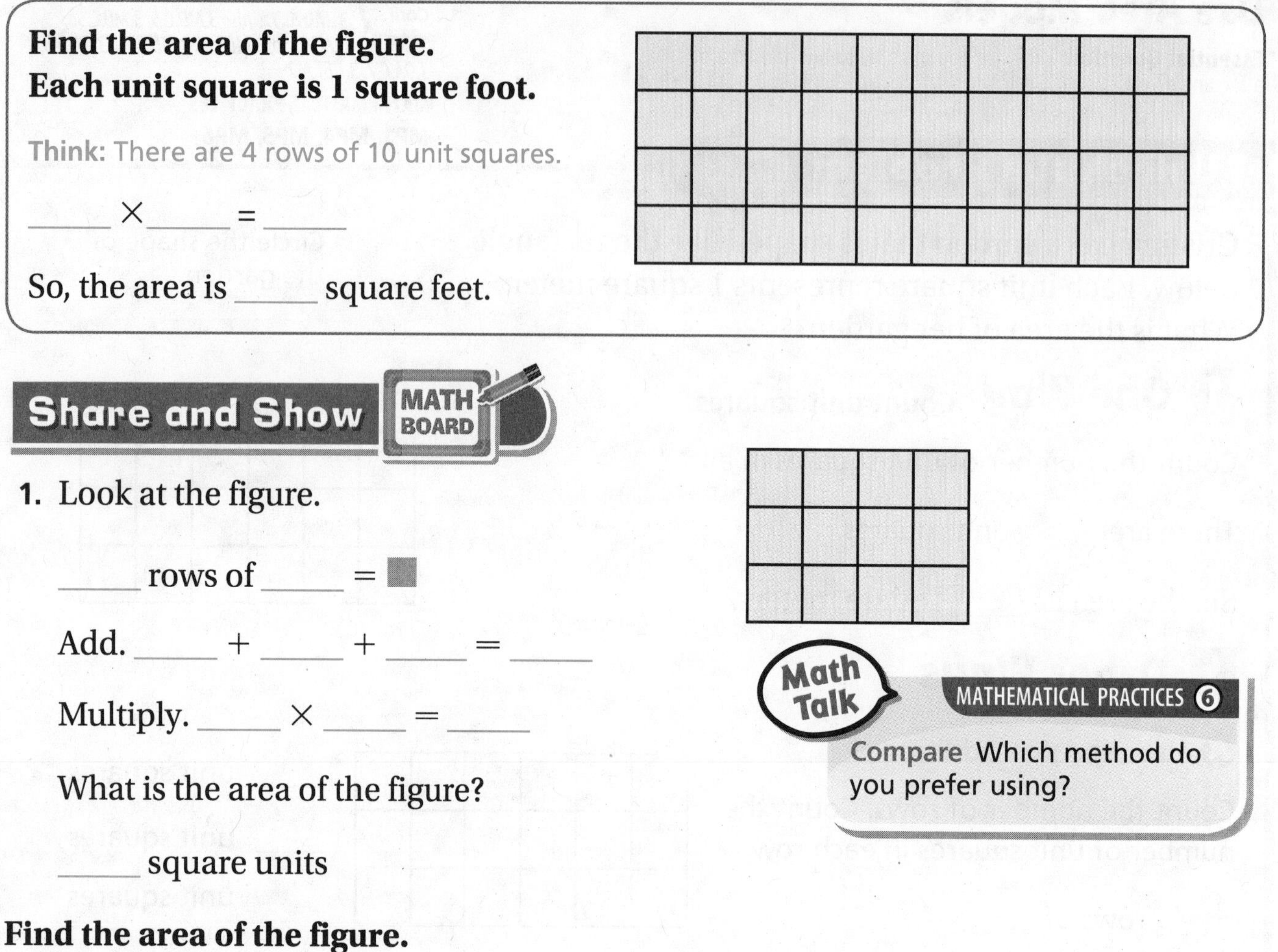

Try This!

Find the area of the figure.
Each unit square is 1 square foot.

Think: There are 4 rows of 10 unit squares.

_____ × _____ = _____

So, the area is _____ square feet.

Share and Show

MATH BOARD

1. Look at the figure.

   _____ rows of _____ = ■

   Add. _____ + _____ + _____ = _____

   Multiply. _____ × _____ = _____

   What is the area of the figure?

   _____ square units

Math Talk

MATHEMATICAL PRACTICES 6

Compare Which method do you prefer using?

Find the area of the figure.
Each unit square is 1 square foot.

2. _____

3. _____

Find the area of the figure.
Each unit square is 1 square meter.

4. _____

5. 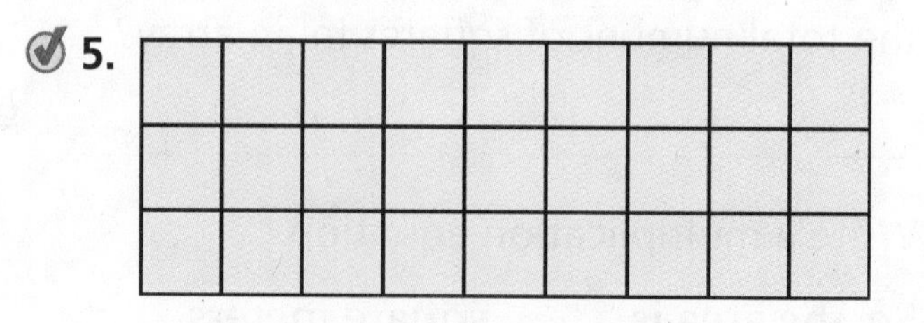

_____

Name ___

On Your Own

Find the area of the figure.
Each unit square is 1 square foot.

6. 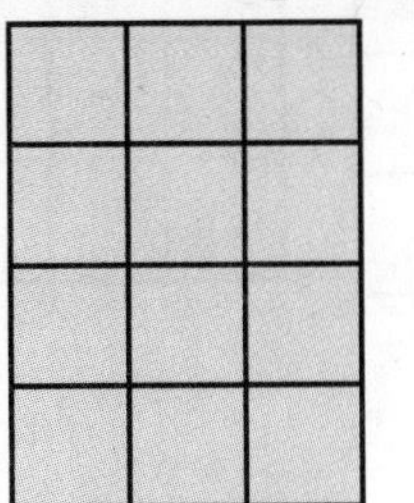

___

7.

___

Find the area of the figure.
Each unit square is 1 square meter.

8.

___

9.

___

10. MATHEMATICAL PRACTICE 4 **Use Diagrams** Draw and shade three rectangles with an area of 24 square units. Then write an addition or multiplication equation for each.

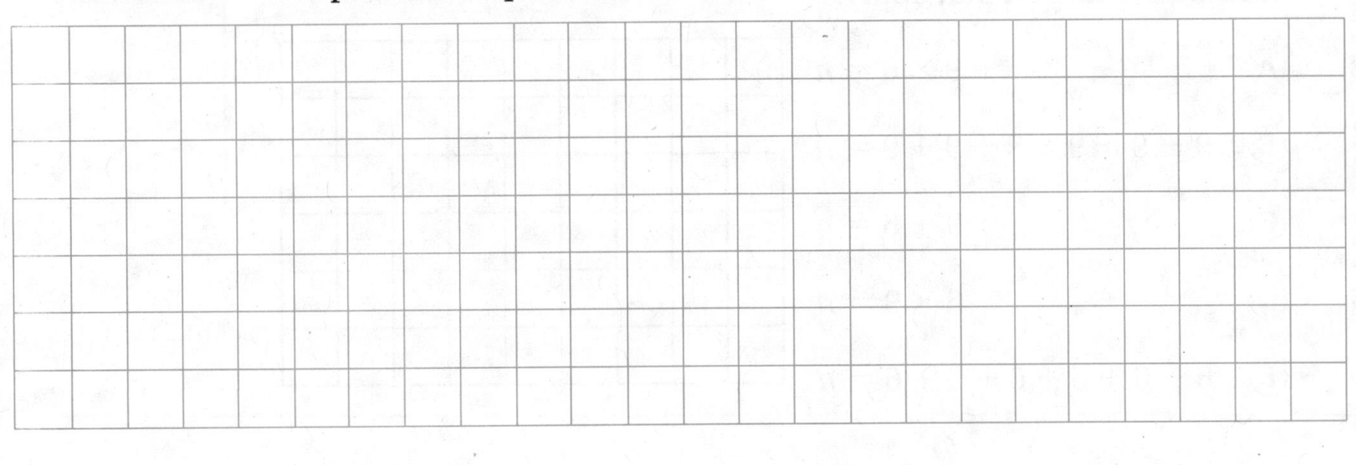

___

___

Problem Solving • Applications Real World

11. GO DEEPER Compare the areas of the two rugs at the right. Each unit square represents 1 square foot. Which rug has the greater area? Explain.

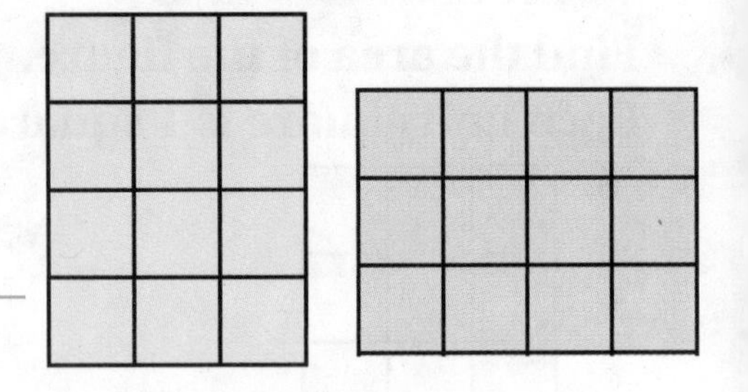

____

____

____

12. THINK SMARTER A tile company tiled a wall using square tiles. A mural is painted in the center. The drawing shows the design. The area of each tile used is 1 square foot.

Write a problem that can be solved by using the drawing. Then solve your problem.

____

____

____

13. THINK SMARTER Colleen drew this rectangle. Select the equation that can be used to find the area of the rectangle. Mark all that apply.

Ⓐ $9 \times 6 = n$

Ⓑ $9 + 9 + 9 + 9 + 9 + 9 = n$

Ⓒ $9 + 6 = n$

Ⓓ $6 \times 9 = n$

Ⓔ $6 + 6 + 6 + 6 + 6 + 6 = n$

Name ____________________

Use Area Models

COMMON CORE STANDARDS—3.MD.C.7, 3.MD.C.7a *Geometric measurement: understand concepts of area and relate area to multiplication and to addition.*

Find the area of each shape. Each unit square is 1 square foot.

1.

There are 3 rows of 8 unit squares.
$3 \times 8 = 24$

24 square feet

2.

____

Find the area of each shape.
Each unit square is 1 square meter.

3.

____

4.

____

5.

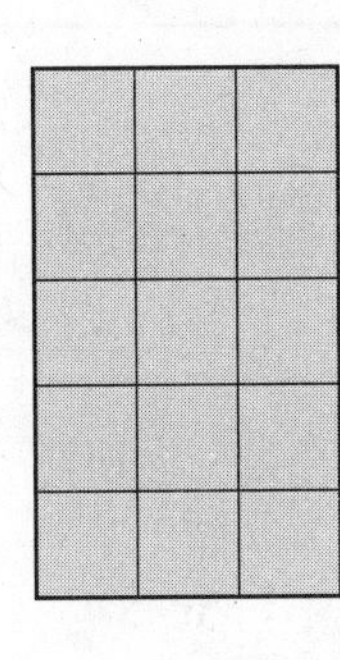

____

Problem Solving Real World

6. Landon made a rug for the hallway. Each unit square is 1 square foot. What is the area of the rug?

____

7. Eva makes a border at the top of a picture frame. Each unit square is 1 square inch. What is the area of the border?

____

8. WRITE Math Describe each of the three methods you can use to find the area of a rectangle.

____

Lesson Check (3.MD.C.7, 3.MD.C.7a)

1. The entrance to an office has a tiled floor. Each square tile is 1 square meter. What is the area of the floor?

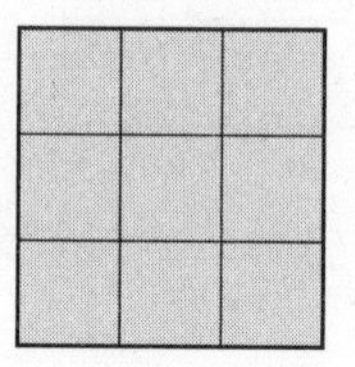

____

2. Ms. Burns buys a new rug. Each unit square is 1 square foot. What is the area of the rug?

____

Spiral Review (3.OA.A.4, 3.NF.A.3d, 3.MD.A.1, 3.MD.D.8)

3. Compare the fractions. Write <, >, or =.

$\frac{1}{3} \bigcirc \frac{2}{3}$

4. Claire bought 6 packs of baseball cards. Each pack had the same number of cards. If Claire bought 48 baseball cards in all, how many cards were in each pack?

____

5. Austin left for school at 7:35 A.M. He arrived at school 15 minutes later. What time did Austin arrive at school?

____

6. Wyatt's room is a rectangle with a perimeter of 40 feet. The width of the room is 8 feet. What is the length of the room?

____

FOR MORE PRACTICE GO TO THE **Personal Math Trainer**

Name ______________________________

Mid-Chapter Checkpoint

Vocabulary

Vocabulary
area
perimeter
square unit

Choose the best term from the box.

1. The distance around a figure is the __________. (p. 625)

2. The measure of the number of unit squares needed to cover a figure with no gaps or overlaps is the ______. (p. 643)

Concepts and Skills

Find the perimeter of the figure. Each unit is 1 centimeter. (3.MD.D.8)

3.

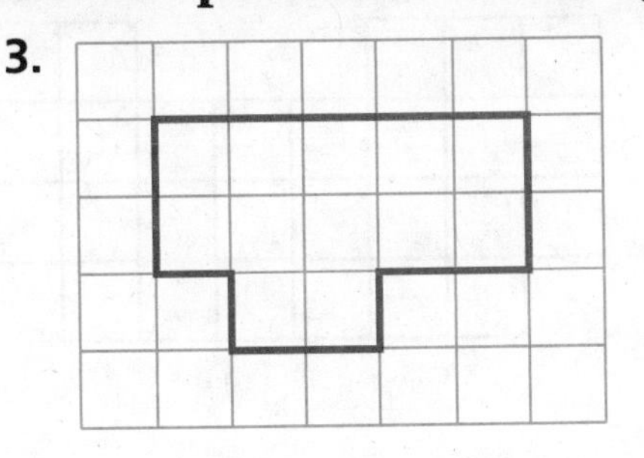

_____ centimeters

4.

_____ centimeters

Find the unknown side lengths. (3.MD.D.8)

5. Perimeter = 33 centimeters

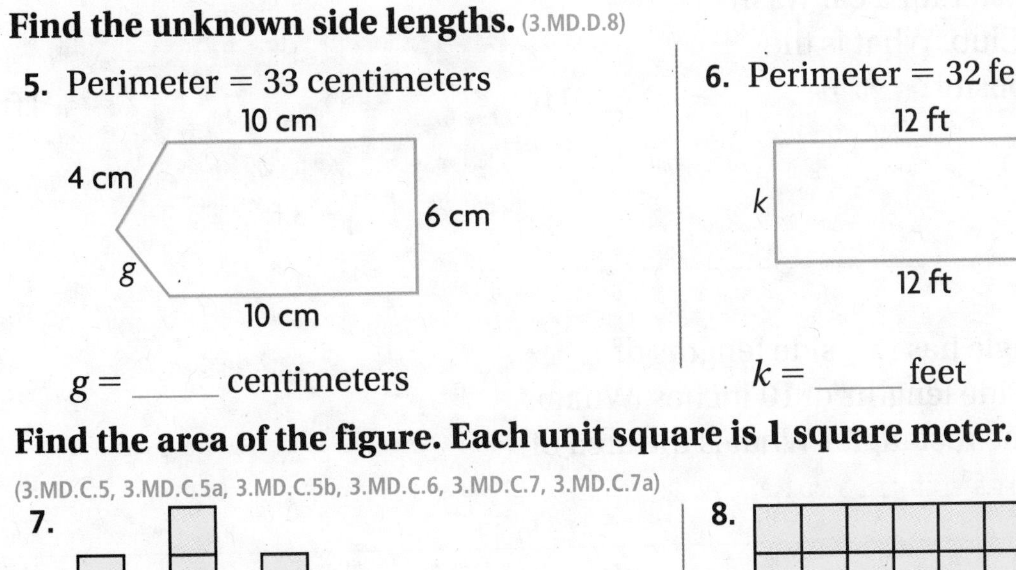

g = _____ centimeters

6. Perimeter = 32 feet

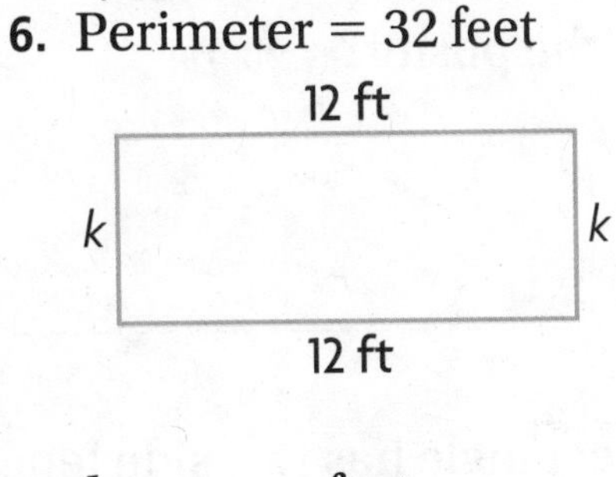

k = _____ feet

Find the area of the figure. Each unit square is 1 square meter.

(3.MD.C.5, 3.MD.C.5a, 3.MD.C.5b, 3.MD.C.6, 3.MD.C.7, 3.MD.C.7a)

7.

_____ square meters

8.

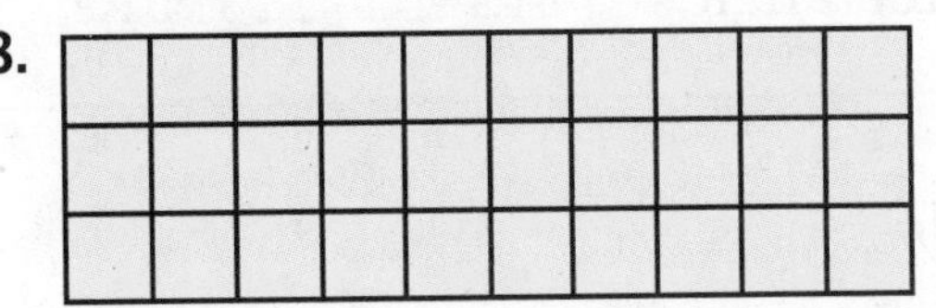

_____ square meters

9. Ramona is making a lid for her rectangular jewelry box. The jewelry box has side lengths of 6 centimeters and 4 centimeters. What is the area of the lid Ramona is making? (3.MD.C.7, 3.MD.C.7a)

6 cm

4 cm

____

10. Adrienne is decorating a square picture frame. She glued 36 inches of ribbon around the edge of the frame. What is the length of each side of the picture frame? (3.MD.D.8)

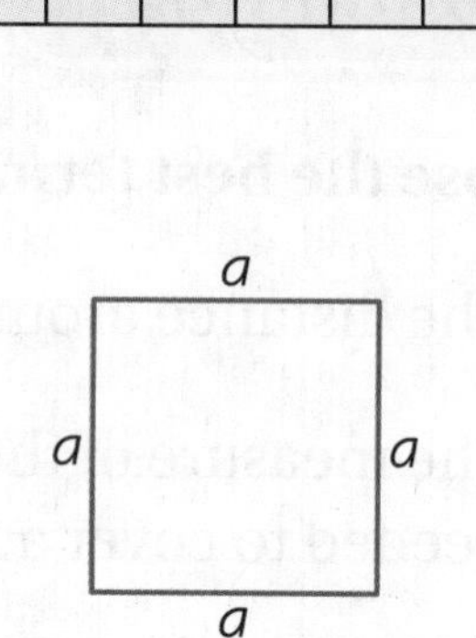

____

11. Margo will sweep a room. A diagram of the floor that she needs to sweep is shown at the right. What is the area of the floor? (3.MD.C.5b, 3.MD.C.6)

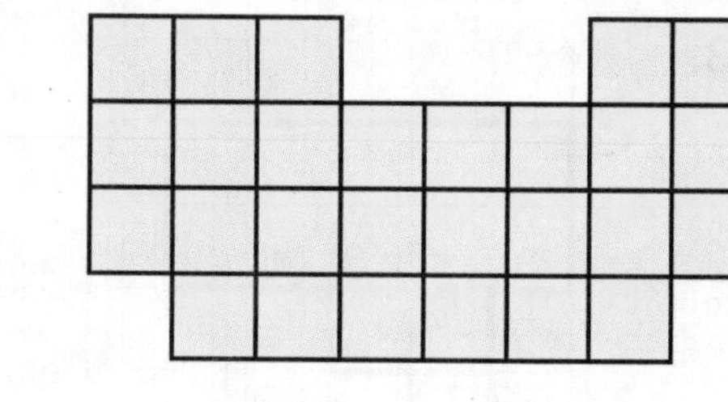

____

12. Jeff is making a poster for a car wash for the Campout Club. What is the perimeter of the poster? (3.MD.D.8)

____

13. GO DEEPER A rectangle has two side lengths of 8 inches and two side lengths of 10 inches. What is the perimeter of the rectangle? What is the area of the rectangle? (3.MD.C.5, 3.MD.C.5a, 3.MD.D.8)

____

Name ______________________________

Problem Solving • Area of Rectangles

Essential Question How can you use the strategy *find a pattern* to solve area problems?

Common Core **Measurement and Data—3.MD.C.7b**
Also 3.OA.A.3, 3.OA.C.7, 3.OA.D.9

MATHEMATICAL PRACTICES
MP1, MP2, MP7

Unlock the Problem (Real World)

Mr. Koi wants to build storage buildings, so he drew plans for the buildings. He wants to know how the areas of the buildings are related. How does the area change from the area of Building *A* to the area of Building *B*? How does the area change from the area of Building *C* to the area of Building *D*?

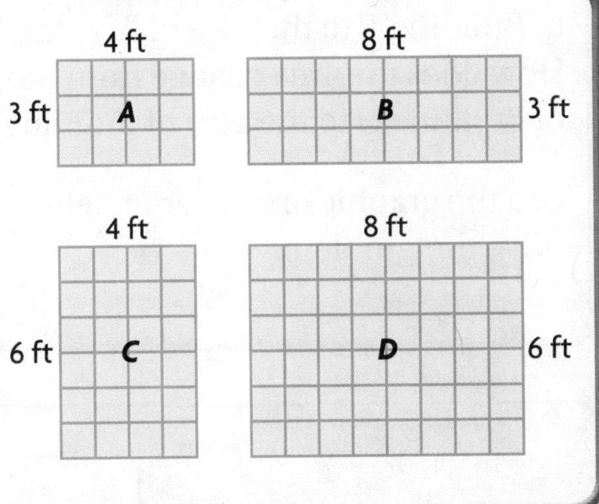

Use the graphic organizer to help you solve the problem.

Read the Problem

What do I need to find?

I need to find how the areas will change from *A* to *B* and from _____ to _____.

What information do I need to use?

I need to use the _____ and _____ of each building to find its area.

How will I use the information?

I will record the areas in a table. Then I will look for a pattern to see how the _____ will change.

Solve the Problem

I will complete the table to find patterns to solve the problem.

	Length	Width	Area		Length	Width	Area
Building *A*	3 ft			Building *C*		4 ft	
Building *B*	3 ft			Building *D*		8 ft	

I see that the lengths will be the same and the widths will be doubled.

The areas will change from _____ to _____ and from _____ to _____.

So, when the lengths are the same and the widths are doubled,

the areas will be __________.

Try Another Problem

Mr. Koi is building more storage buildings. He wants to know how the areas of the buildings are related. How does the area change from the area of Building *E* to the area of Building *F*? How does the area change from the area of Building *G* to the area of Building *H*?

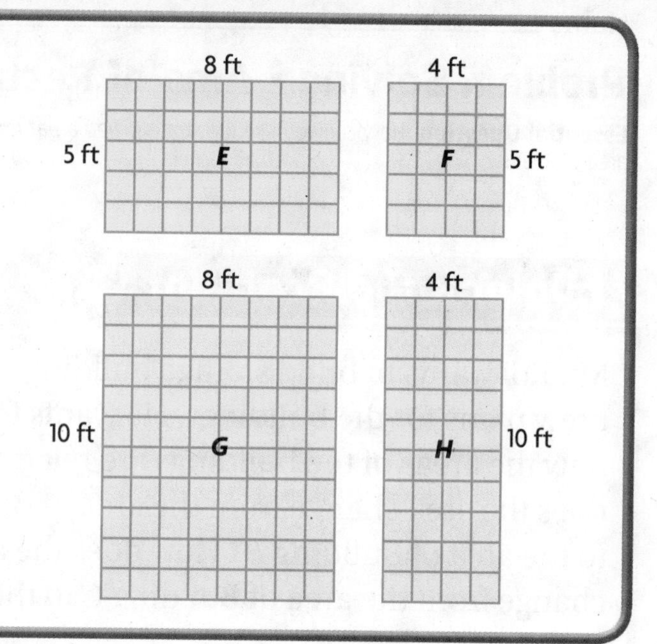

Use the graphic organizer to help you solve the problem.

Read the Problem

What do I need to find?	What information do I need to use?	How will I use the information?

Solve the Problem

	Length	Width	Area		Length	Width	Area
Building *E*				Building *G*			
Building *F*				Building *H*			

- How did your table help you find a pattern?

____

____

____

MATHEMATICAL PRACTICES 2

Reason Abstractly What if the length of both sides is doubled? How would the areas change?

Name ________________________________

Share and Show 

Use the table for 1–2.

Swimming Pool Sizes			
Pool	Length (in feet)	Width (in feet)	Area (in square feet)
A	8	20	
B	8	30	
C	8	40	
D	8	50	

✓ **1.** Many pools come in rectangular shapes. How do the areas of the swimming pools change when the widths change?

First, complete the table by finding the area of each pool.

Think: I can find the area by multiplying the length and the width.

Then, find a pattern of how the lengths change and how the widths change.

The ________ stays the same. The widths

______________________.

Last, describe a pattern of how the area changes.

The areas ________ by ____ square feet.

✓ **2.** What if the length of each pool was 16 feet? Explain how the areas would change.

____

On Your Own

3. MATHEMATICAL PRACTICE 7 **Look for a Pattern** If the length of each pool in the table is 20 feet, and the widths change from 5, to 6, to 7, and to 8 feet, describe the pattern of the areas.

____

____

4. MATHEMATICAL PRACTICE 1 **Analyze Relationships** Jacob has a rectangular garden with an area of 56 square feet. The length of the garden is 8 feet. What is the width of the garden?

____

5. GO DEEPER A diagram of Paula's bedroom is at the right. Her bedroom is in the shape of a rectangle. Write the measurements for the other sides. What is the perimeter of the room? (Hint: The two pairs of opposite sides are equal lengths.)

17 ft

12

____

6. THINK SMARTER Elizabeth built a sandbox that is 4 feet long and 4 feet wide. She also built a flower garden that is 4 feet long and 6 feet wide and a vegetable garden that is 4 feet long and 8 feet wide. How do the areas change?

____

____

____

____

7. THINK SMARTER Find the pattern and complete the chart.

Total Area (in square feet)	50	60	70	80	
Length (in feet)	10	10		10	
Width (in feet)	5	6	7		

How can you use the chart to find the length and width of a figure with an area of 100 square feet?

____

____

Name ______________________

Problem Solving • Area of Rectangles

COMMON CORE STANDARD—3.MD.C.7b
Geometric measurement: understand concepts of area and relate area to multiplication and to addition.

Use the information for 1–3.

An artist makes rectangular murals in different sizes. Below are the available sizes. Each unit square is 1 square meter.

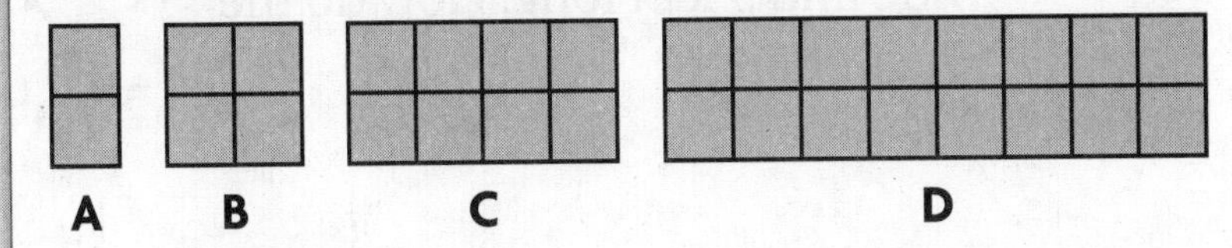

1. Complete the table to find the area of each mural.

Mural	Length (in meters)	Width (in meters)	Area (in square meters)
A	2	1	2
B	2	2	4
C	2		
D	2		

2. Find and describe a pattern of how the length changes and how the width changes for murals A through D.

____

____

3. How do the areas of the murals change when the width changes?

____

4. WRITE ▸*Math* Write and solve an area problem that illustrates the use of the *find a pattern* strategy.

____

____

Lesson Check (3.MD.C.7b)

1. Lauren drew the designs below. Each unit square is 1 square centimeter. If the pattern continues, what will be the area of the fourth figure?

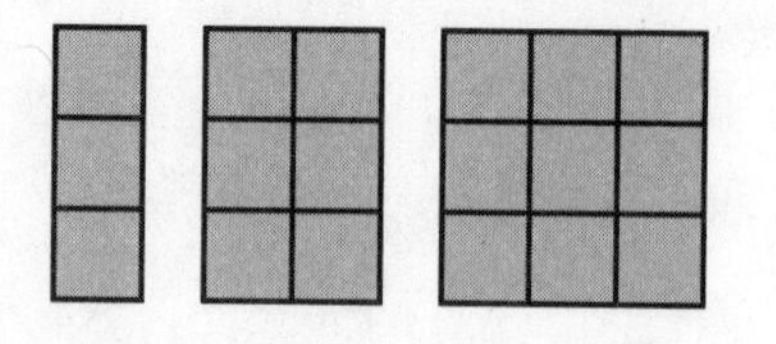

2. Henry built one garden that is 3 feet wide and 3 feet long. He also built a garden that is 3 feet wide and 6 feet long, and a garden that is 3 feet wide and 9 feet long. How do the areas change?

Spiral Review (3.OA.A.3, 3.NBT.A.3, 3.NF.A.1, 3.MD.C.5b, 3.MD.C.6)

3. Joe, Jim, and Jack share 27 football cards equally. How many cards does each boy get?

4. Nita uses $\frac{1}{3}$ of a carton of 12 eggs. How many eggs does she use?

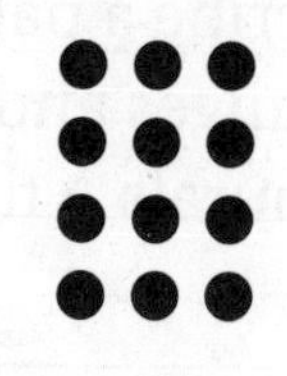

5. Brenda made 8 necklaces. Each necklace has 10 large beads. How many large beads did Brenda use to make the necklaces?

6. Neal is tiling his kitchen floor. Each square tile is 1 square foot. Neal uses 6 rows of tiles with 9 tiles in each row. What is the area of the floor?

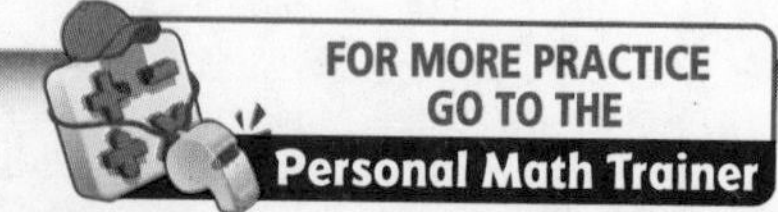

Name ______________________________

Area of Combined Rectangles

Essential Question How can you break apart a figure to find the area?

Common Core **Measurement and Data—3.MD.C.7c, 3.MD.C.7d**
Also 3.MD.C.5, 3.MD.C.5a, 3.MD.C.5b, 3.MD.C.7b, 3.OA.A.3, 3.OA.B.5, 3.OA.C.7, 3.NBT.A.2
MATHEMATICAL PRACTICES
MP1, MP4, MP6

Unlock the Problem

Real World · Hands On

Anna's rug has side lengths of 4 feet and 9 feet. What is the area of Anna's rug?

Remember

You can use the Distributive Property to break apart an array.

$3 \times 3 = 3 \times (2 + 1)$

Activity **Materials** ■ square tiles

STEP 1 Use square tiles to model 4×9.

STEP 2 Draw a rectangle on the grid paper to show your model.

STEP 3 Draw a vertical line to break apart the model to make two smaller rectangles.

The side length 9 is broken into ____ plus ____.

STEP 4 Find the area of each of the two smaller rectangles.

Rectangle 1: ____ × ____ = ____

Rectangle 2: ____ × ____ = ____

STEP 5 Add the products to find the total area.

____ + ____ = ____ square feet

STEP 6 Check your answer by counting the number of square feet.

____ square feet

So, the area of Anna's rug is ____ square feet.

Math Talk MATHEMATICAL PRACTICES 6

Compare Did you draw a line in the same place as your classmates? Explain why you found the same total area.

CONNECT Using the Distributive Property, you found that you could break apart a rectangle into smaller rectangles, and add the area of each smaller rectangle to find the total area.

How can you break apart this figure into rectangles to find its area?

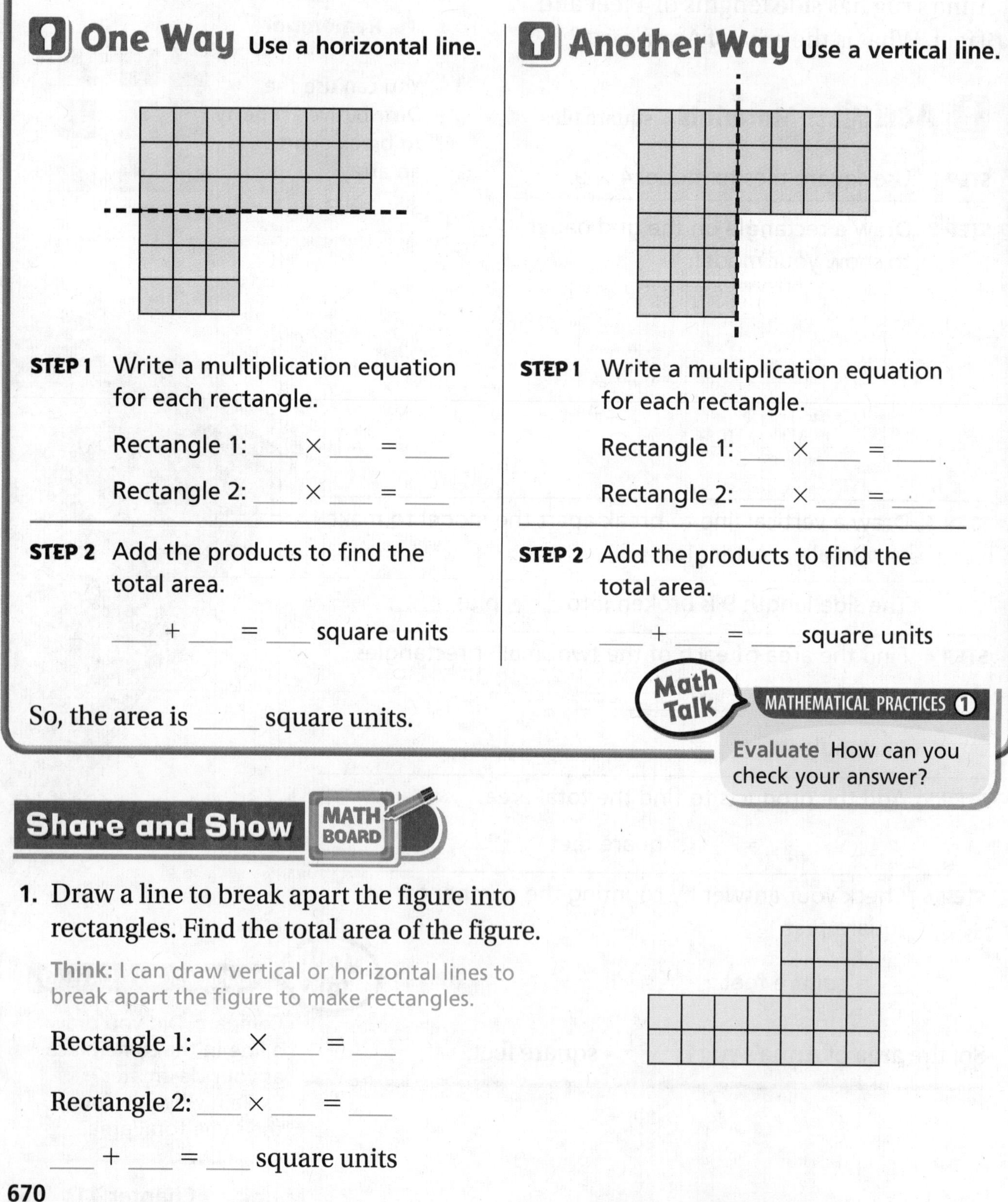

One Way Use a horizontal line.

STEP 1 Write a multiplication equation for each rectangle.

Rectangle 1: ___ × ___ = ___

Rectangle 2: ___ × ___ = ___

STEP 2 Add the products to find the total area.

___ + ___ = ___ square units

Another Way Use a vertical line.

STEP 1 Write a multiplication equation for each rectangle.

Rectangle 1: ___ × ___ = ___

Rectangle 2: ___ × ___ = ___

STEP 2 Add the products to find the total area.

___ + ___ = ___ square units

So, the area is _____ square units.

Math Talk MATHEMATICAL PRACTICES 1

Evaluate How can you check your answer?

Share and Show MATH BOARD

1. Draw a line to break apart the figure into rectangles. Find the total area of the figure.

   Think: I can draw vertical or horizontal lines to break apart the figure to make rectangles.

   Rectangle 1: ___ × ___ = ___

   Rectangle 2: ___ × ___ = ___

   ___ + ___ = ___ square units

Name ________________

Use the Distributive Property to find the area. Show your multiplication and addition equations.

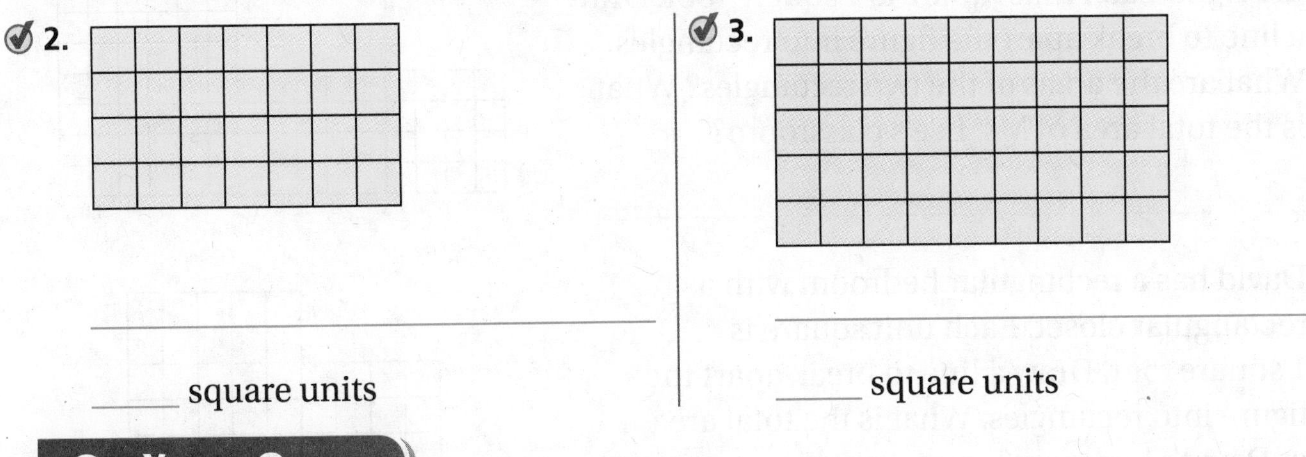

2. ______________________

____ square units

3. ______________________

____ square units

On Your Own

Use the Distributive Property to find the area. Show your multiplication and addition equations.

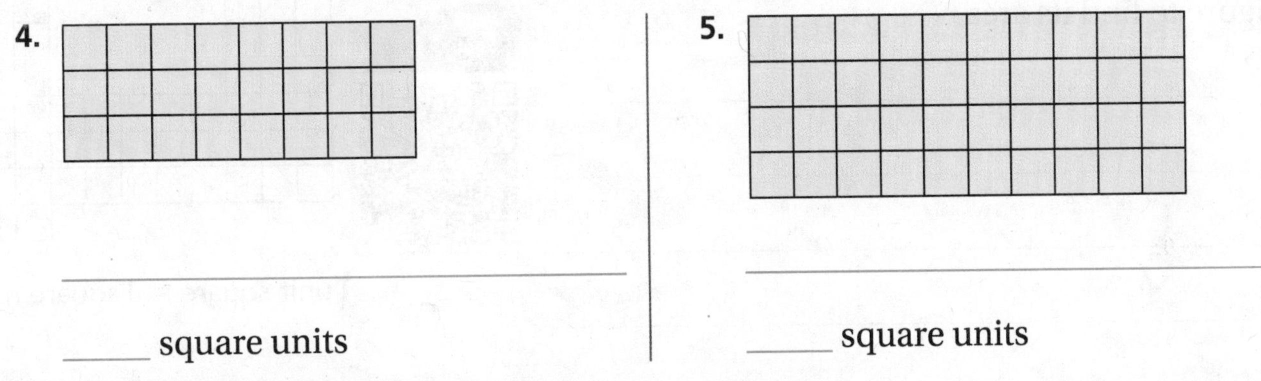

4. ______________________

____ square units

5. ______________________

____ square units

Draw a line to break apart the figure into rectangles. Find the area of the figure.

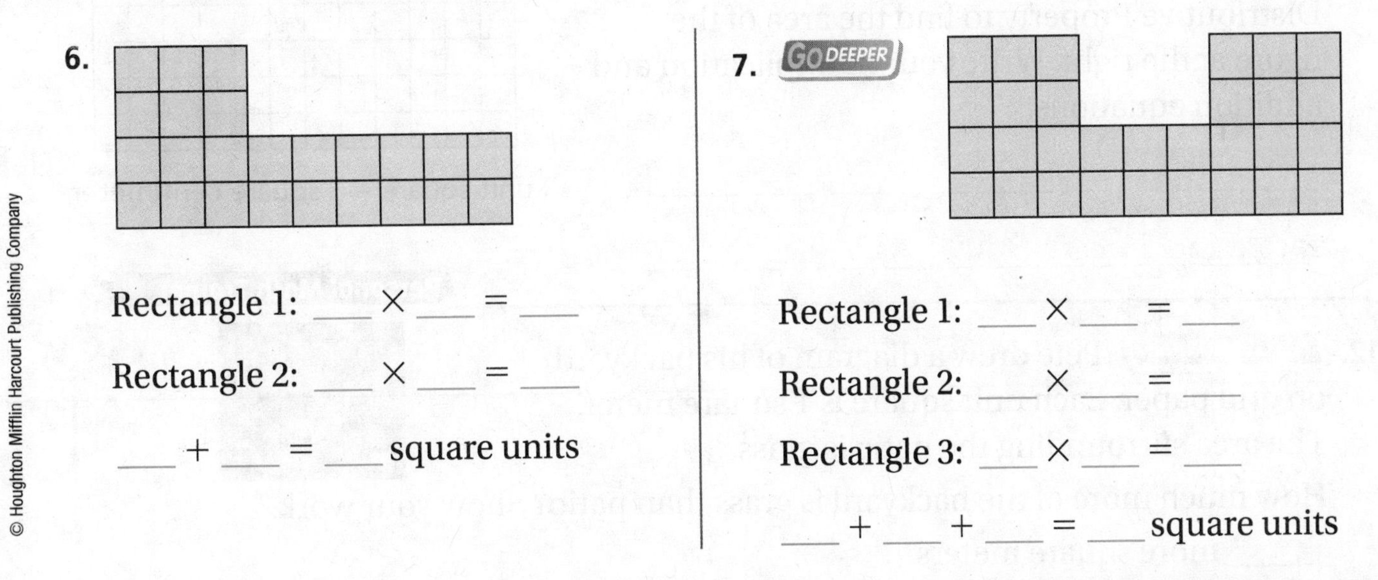

6.

Rectangle 1: ___ × ___ = ___

Rectangle 2: ___ × ___ = ___

___ + ___ = ___ square units

7. GO DEEPER

Rectangle 1: ___ × ___ = ___

Rectangle 2: ___ × ___ = ___

Rectangle 3: ___ × ___ = ___

___ + ___ + ___ = ___ square units

Problem Solving • Applications Real World

8. GO DEEPER A model of Ms. Lee's classroom is at the right. Each unit square is 1 square foot. Draw a line to break apart the figure into rectangles. What are the areas of the two rectangles? What is the total area of Ms. Lee's classroom?

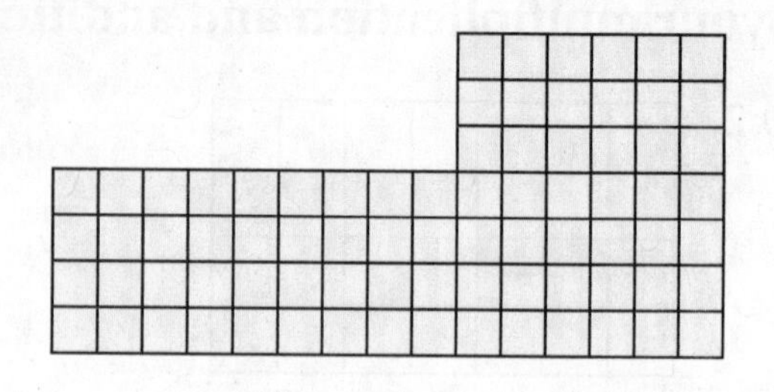

____

9. David has a rectangular bedroom with a rectangular closet. Each unit square is 1 square foot. Draw a line to break apart the figure into rectangles. What is the total area of David's bedroom?

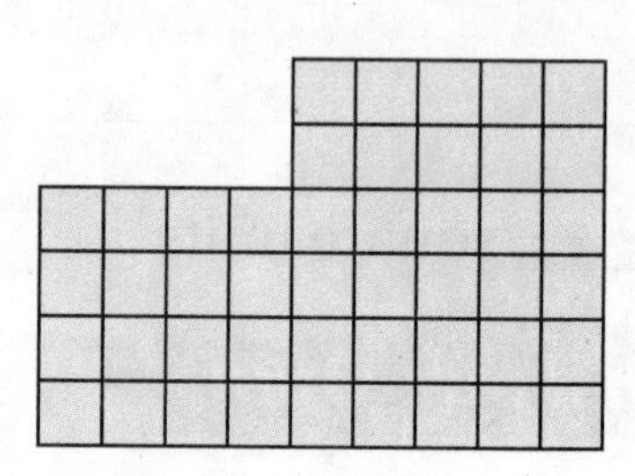

____

10. THINK SMARTER **Explain** how to break apart the figure to find its area.

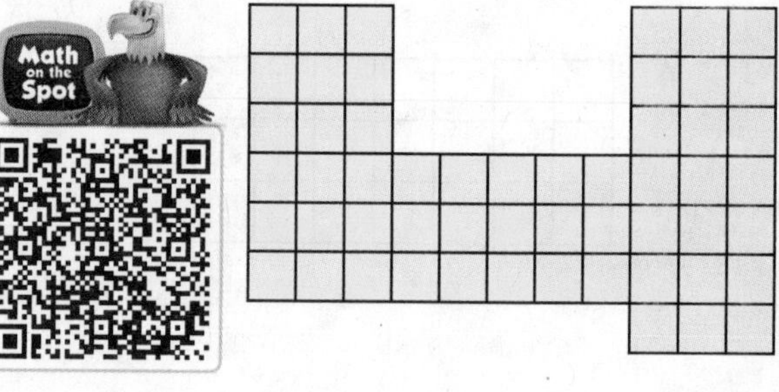

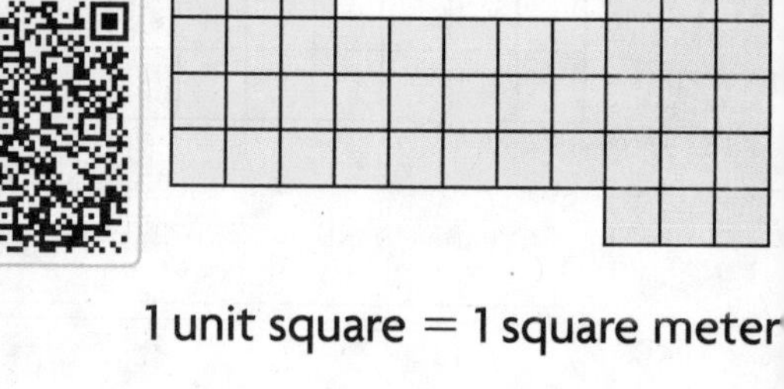

1 unit square = 1 square meter

____

____

____

____

____

11. MATHEMATICAL PRACTICE 4 **Interpret a Result** Use the Distributive Property to find the area of the figure at the right. Write your multiplication and addition equations.

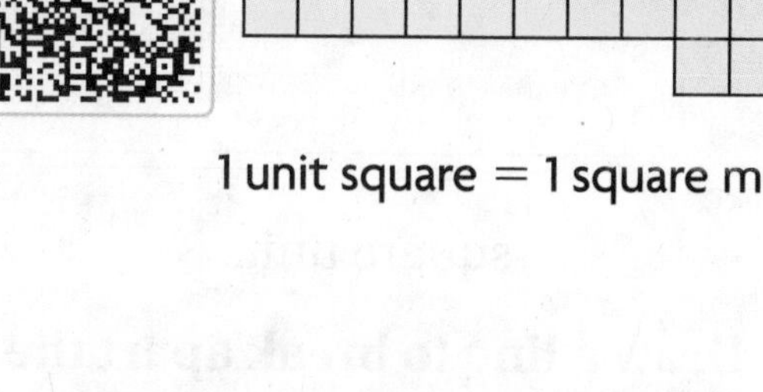

1 unit square = 1 square centimeter

____

____

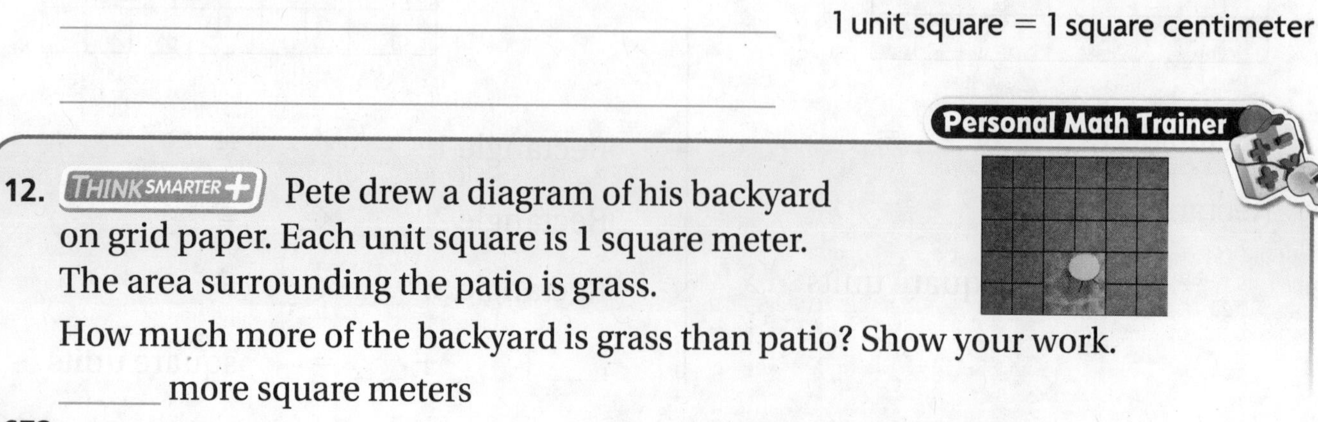

Personal Math Trainer

12. THINK SMARTER + Pete drew a diagram of his backyard on grid paper. Each unit square is 1 square meter. The area surrounding the patio is grass.

How much more of the backyard is grass than patio? Show your work.

______ more square meters

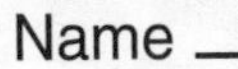
Name ______________________

Area of Combined Rectangles

COMMON CORE STANDARDS—3.MD.C.7c, 3.MD.C.7d *Geometric measurement: understand concepts of area and relate area to multiplication and to addition.*

Use the Distributive Property to find the area. Show your multiplication and addition equations.

1. 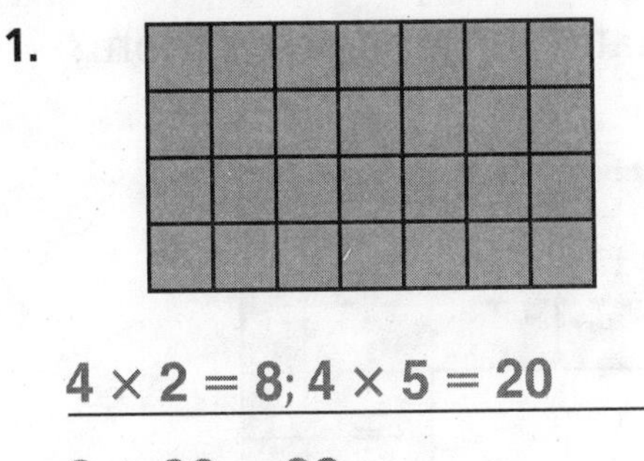

$4 \times 2 = 8$; $4 \times 5 = 20$

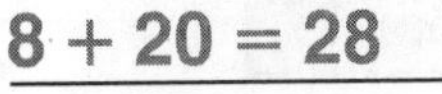
$8 + 20 = 28$

28 square units

2. 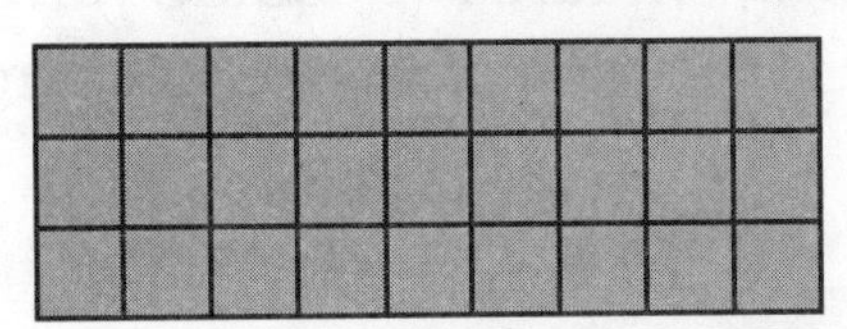

____

____

_____ square units

Draw a line to break apart the shape into rectangles. Find the area of the shape.

3.

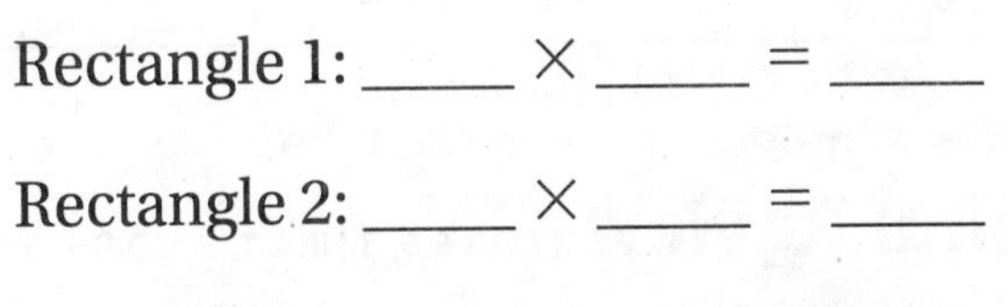
Rectangle 1: _____ × _____ = _____

Rectangle 2: _____ × _____ = _____

_____ + _____ = _____ square units

Problem Solving Real World

A diagram of Frank's room is at right. Each unit square is 1 square foot.

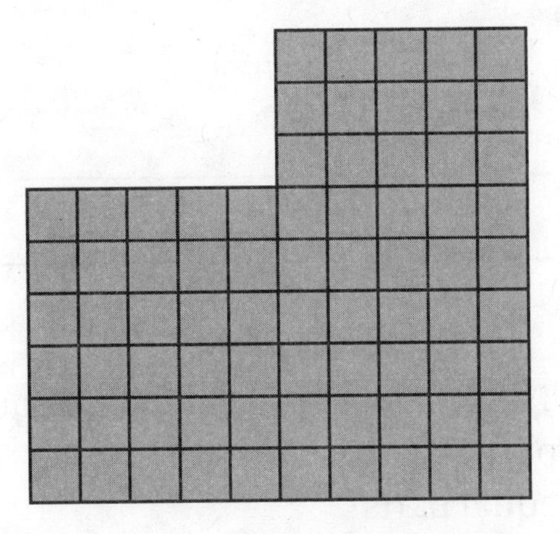

4. Draw a line to divide the shape of Frank's room into rectangles.

5. What is the total area of Frank's room?

   _____ square feet

6. 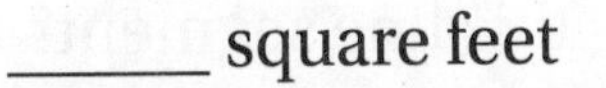 WRITE Math Draw a figure that is not a rectangle and find its area. Use grid paper and show each step.

____

____

Lesson Check (3.MD.C.7c, 3.MD.C.7d)

1. The diagram shows Ben's backyard. Each unit square is 1 square yard. What is the area of Ben's backyard?

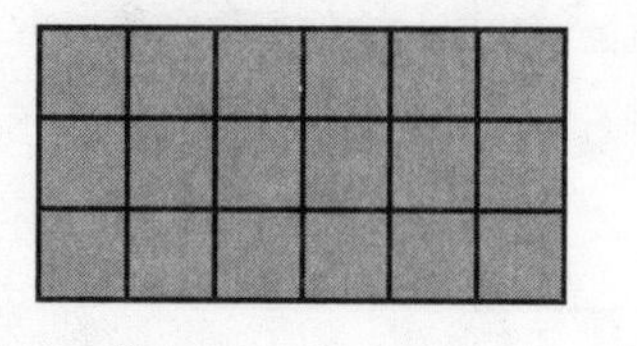

2. The diagram shows a room in an art gallery. Each unit square is 1 square meter. What is the area of the room?

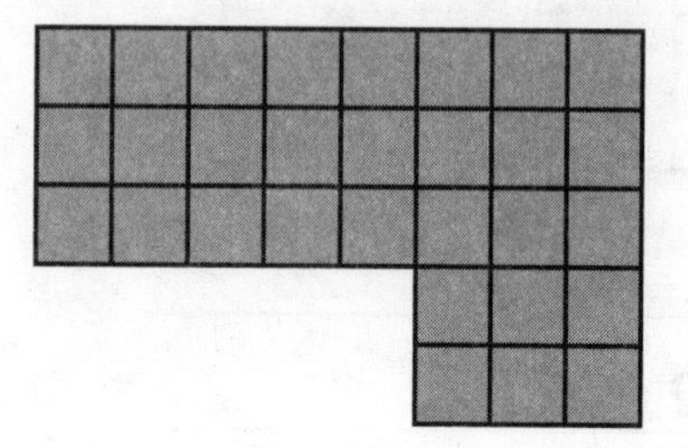

Spiral Review (3.OA.B.6, 3.NF.A.1, 3.MD.B.4, 3.MD.D.8)

3. Naomi needs to solve $28 \div 7 = \blacksquare$. What related multiplication fact can she use to find the unknown number?

4. Karen drew a triangle with side lengths 3 centimeters, 4 centimeters, and 5 centimeters. What is the perimeter of the triangle?

5. The rectangle is divided into equal parts. What is the name of the equal parts?

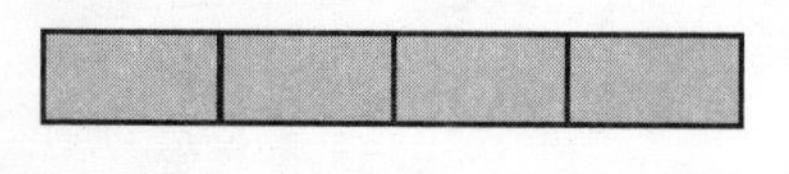

6. Use an inch ruler. To the nearest half inch, how long is this line segment?

Name ______________________

Same Perimeter, Different Areas

Essential Question How can you use area to compare rectangles with the same perimeter?

Common Core **Measurement and Data—3.MD.D.8**
Also 3.MD.C.5, 3.MD.C.5a, 3.MD.C.5b, 3.MD.C.7b, 3.OA.A.3, 3.OA.C.7, 3.NBT.A.2

MATHEMATICAL PRACTICES
MP2, MP3, MP4, MP6

Unlock the Problem (Real World)

Hands On

Toby has 12 feet of boards to put around a rectangular sandbox. How long should he make each side so that the area of the sandbox is as large as possible?

- What is the greatest perimeter Toby can make for his sandbox?

____

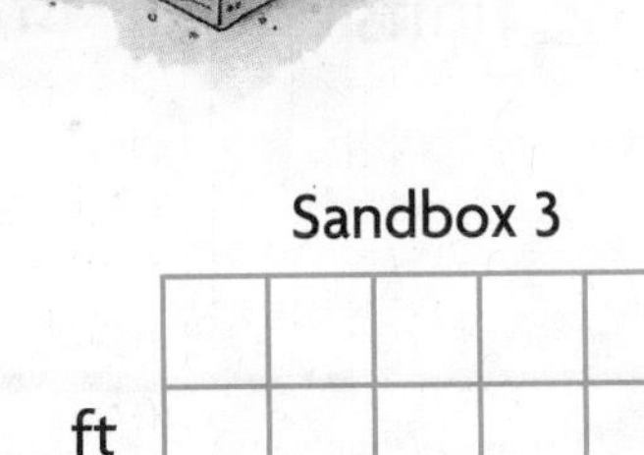

Activity

Materials ■ square tiles

Use square tiles to make all the rectangles you can that have a perimeter of 12 units. Draw and label the sandboxes. Then find the area of each.

Sandbox 1

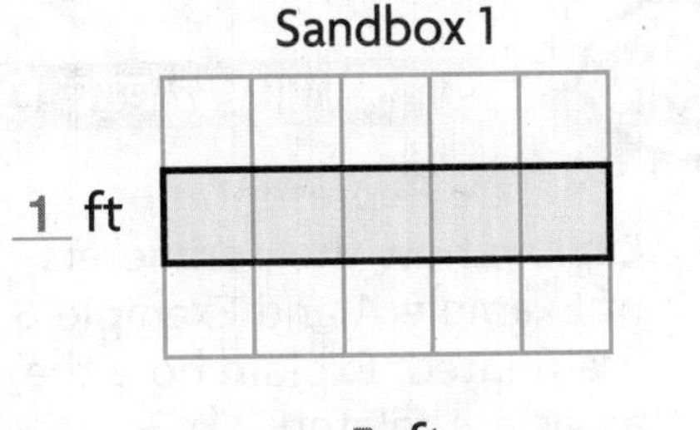

1 ft

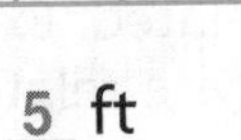

5 ft

Sandbox 2

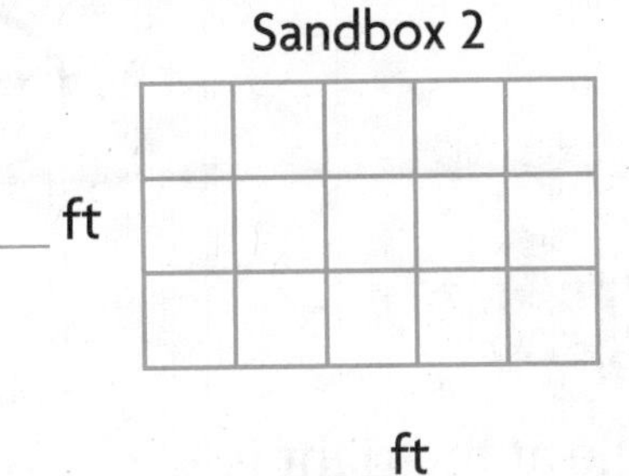

___ ft

___ ft

Sandbox 3

___ ft

___ ft

Find the perimeter and area of each rectangle.

	Perimeter	Area
Sandbox 1	5 + 1 + 5 + 1 = 12 feet	1 × 5 = ___ square feet
Sandbox 2	___ + ___ + ___ + ___ = ___ feet	___ × ___ = ___ square feet
Sandbox 3	___ + ___ + ___ + ___ = ___ feet	___ × ___ = ___ square feet

The area of Sandbox ____ is the greatest.

So, Toby should build a sandbox that is

____ feet wide and ____ feet long.

Math Talk MATHEMATICAL PRACTICES 6

Compare How are the sandboxes alike? How are the sandboxes different?

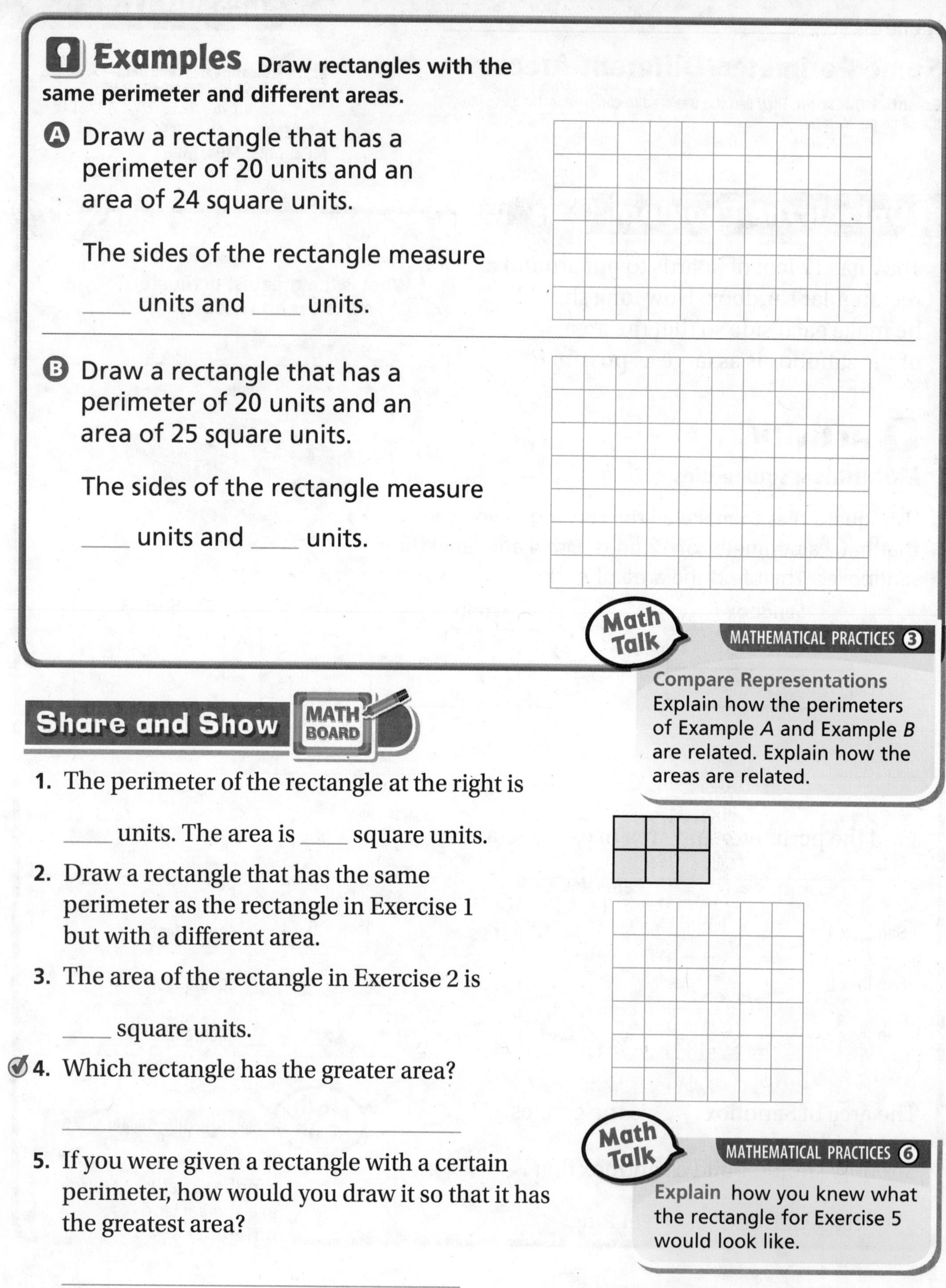

Examples Draw rectangles with the same perimeter and different areas.

A Draw a rectangle that has a perimeter of 20 units and an area of 24 square units.

The sides of the rectangle measure

____ units and ____ units.

B Draw a rectangle that has a perimeter of 20 units and an area of 25 square units.

The sides of the rectangle measure

____ units and ____ units.

Math Talk MATHEMATICAL PRACTICES 3

Compare Representations Explain how the perimeters of Example *A* and Example *B* are related. Explain how the areas are related.

Share and Show MATH BOARD

1. The perimeter of the rectangle at the right is

   ____ units. The area is ____ square units.

2. Draw a rectangle that has the same perimeter as the rectangle in Exercise 1 but with a different area.

3. The area of the rectangle in Exercise 2 is

   ____ square units.

4. Which rectangle has the greater area?

   ____

5. If you were given a rectangle with a certain perimeter, how would you draw it so that it has the greatest area?

   ____

Math Talk MATHEMATICAL PRACTICES 6

Explain how you knew what the rectangle for Exercise 5 would look like.

Name ______________________________

Find the perimeter and the area. Tell which rectangle has a greater area.

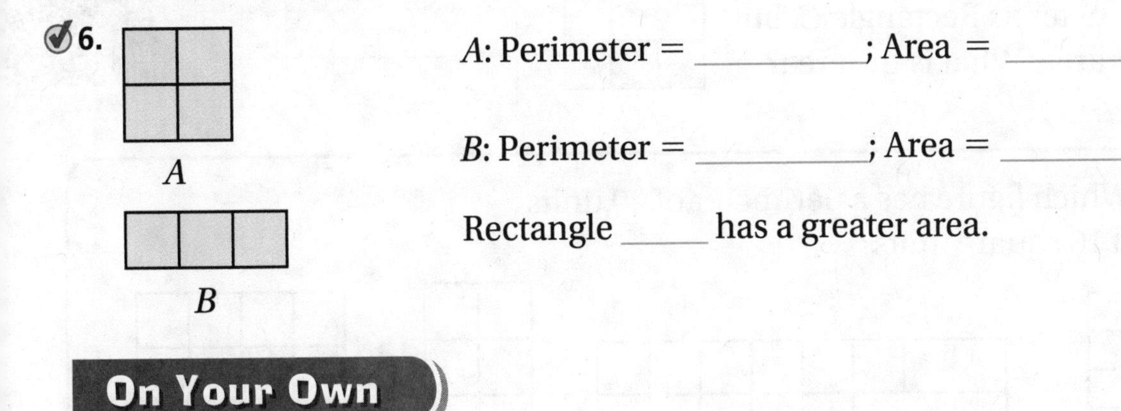

6. A: Perimeter = __________; Area = __________

B: Perimeter = __________; Area = __________

Rectangle ______ has a greater area.

On Your Own

Find the perimeter and the area. Tell which rectangle has a greater area.

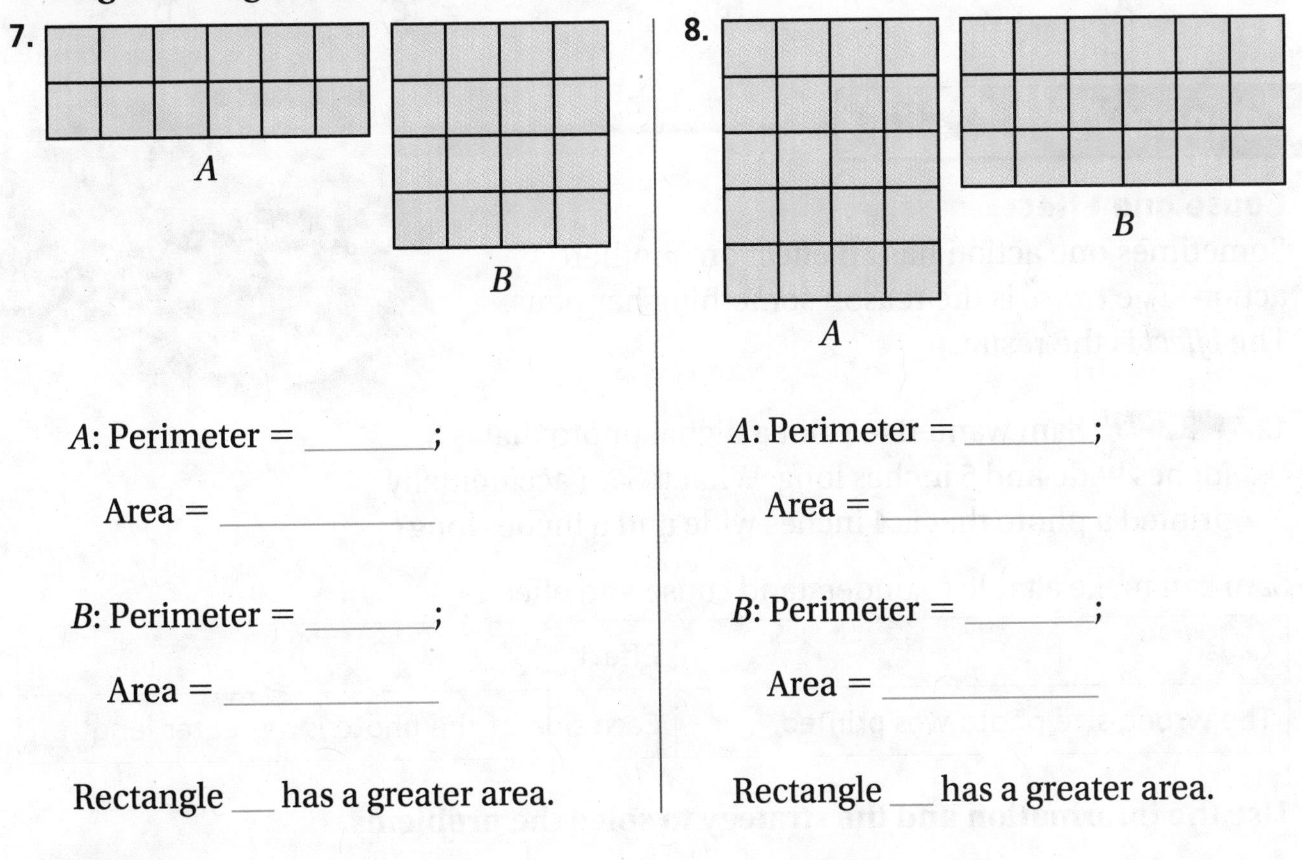

7. A: Perimeter = __________;

Area = __________

B: Perimeter = __________;

Area = __________

Rectangle ___ has a greater area.

8. A: Perimeter = __________;

Area = __________

B: Perimeter = __________;

Area = __________

Rectangle ___ has a greater area.

9. MATHEMATICAL PRACTICE 6 **Use Math Vocabulary** Todd's flower garden is 4 feet wide and 8 feet long. If the answer is 32 square feet, what is the question?

____

Problem Solving • Applications Real World

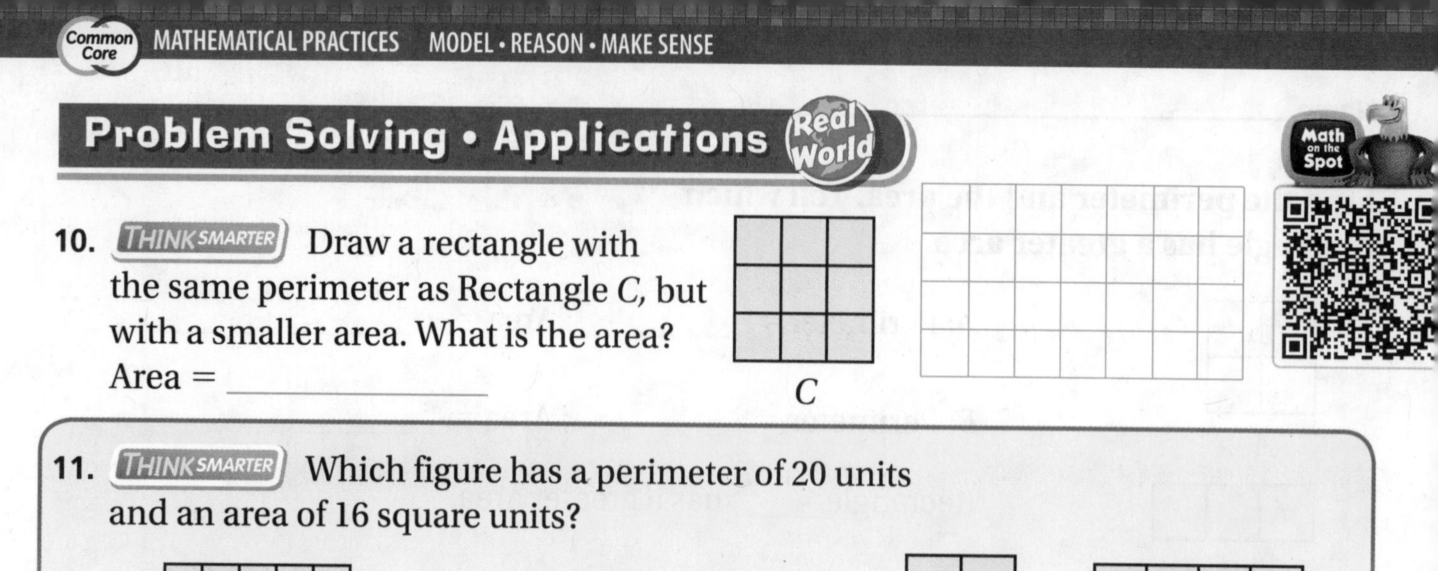

10. THINK SMARTER Draw a rectangle with the same perimeter as Rectangle *C*, but with a smaller area. What is the area?
Area = ____________

11. THINK SMARTER Which figure has a perimeter of 20 units and an area of 16 square units?

Ⓐ Ⓑ Ⓒ Ⓓ

Connect to Reading

Cause and Effect

Sometimes one action has an effect on another action. The *cause* is the reason something happens. The *effect* is the result.

12. GO DEEPER Sam wanted to print a digital photo that is 3 inches wide and 5 inches long. What if Sam accidentally printed a photo that is 4 inches wide and 6 inches long?

Sam can make a table to understand cause and effect.

Cause	Effect
The wrong size photo was printed.	Each side of the photo is a greater length.

Use the information and the strategy to solve the problems.

a. What effect did the mistake have on the perimeter of the photo?

____

b. What effect did the mistake have on the area of the photo?

____

Name ____________________

Same Perimeter, Different Areas

COMMON CORE STANDARD—3.MD.D.8
Geometric measurement: recognize perimeter as an attribute of plane figures and distinguish between linear and area measures.

Find the perimeter and the area.
Tell which rectangle has a greater area.

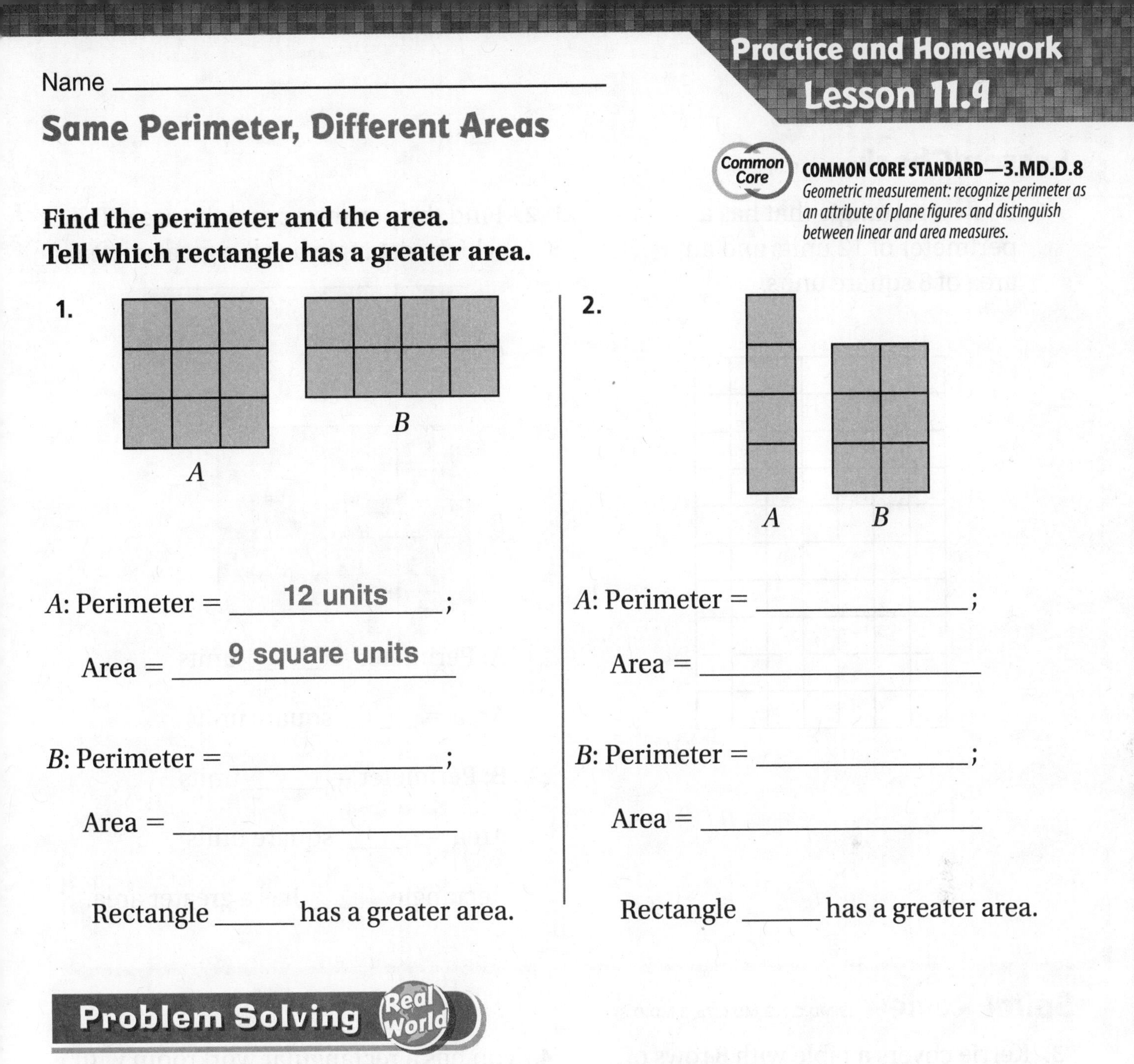

1.

A: Perimeter = 12 units;

Area = 9 square units

B: Perimeter = ______________;

Area = ______________

Rectangle ______ has a greater area.

2.

A: Perimeter = ______________;

Area = ______________

B: Perimeter = ______________;

Area = ______________

Rectangle ______ has a greater area.

Problem Solving Real World

3. Tara's and Jody's bedrooms are shaped like rectangles. Tara's bedroom is 9 feet long and 8 feet wide. Jody's bedroom is 7 feet long and 10 feet wide. Whose bedroom has the greater area? **Explain**.

____

____

4. WRITE Math Draw three examples of rectangles that have the same perimeter, but different areas. Note which of the areas is greatest and which is the least.

____

____

Lesson Check (3.MD.D.8)

1. Draw a rectangle that has a perimeter of 12 units and an area of 8 square units.

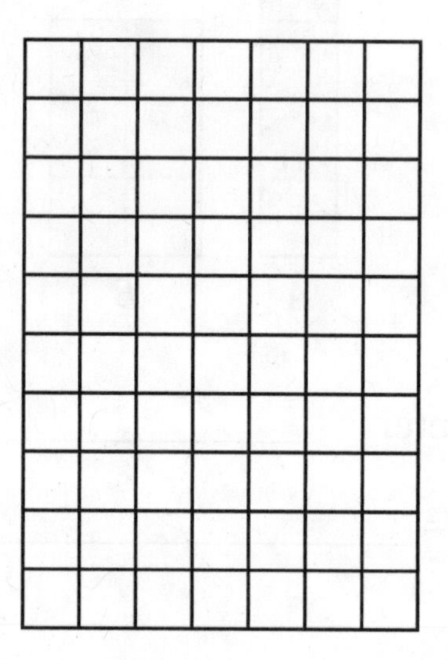

2. Find the perimeter and the area. Tell which rectangle has the greater area.

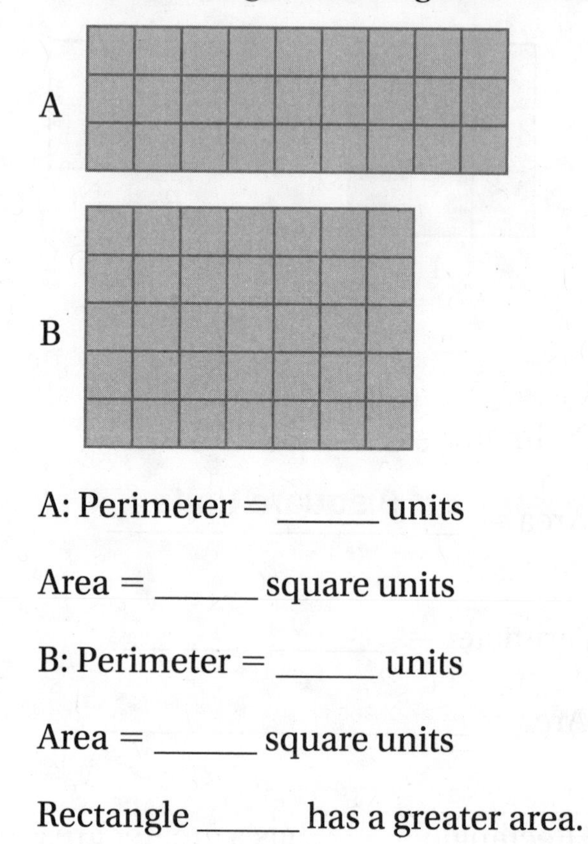

A: Perimeter = ______ units

Area = ______ square units

B: Perimeter = ______ units

Area = ______ square units

Rectangle ______ has a greater area.

Spiral Review (3.MD.C.7, 3.MD.C.7a, 3.MD.D.8)

3. Kerrie covers a table with 8 rows of square tiles. There are 7 tiles in each row. What is the area that Kerrie covers in square units?

____

4. Von has a rectangular workroom with a perimeter of 26 feet. The length of the workroom is 6 feet. What is the width of Von's workroom?

____

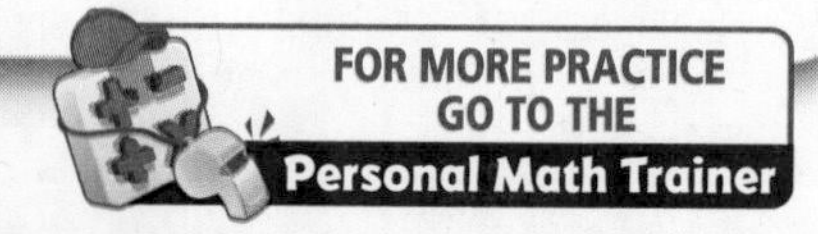

Name ______________________________

Same Area, Different Perimeters

Essential Question How can you use perimeter to compare rectangles with the same area?

Common Core **Measurement and Data—3.MD.D.8**
Also 3.MD.C.5, 3.MD.C.5a, 3.MD.C.5b, 3.MD.C.7b, 3.OA.A.3, 3.OA.C.7, 3.NBT.A.2
MATHEMATICAL PRACTICES
MP2, MP3, MP4, MP6

Unlock the Problem

Real World

Hands On

Marcy is making a rectangular pen to hold her rabbits. The area of the pen should be 16 square meters with side lengths that are whole numbers. What is the least amount of fencing she needs?

- What does the least amount of fencing represent?

____

Activity **Materials** ■ square tiles

Use 16 square tiles to make rectangles. Make as many different rectangles as you can with 16 tiles. Record the rectangles on the grid, write the multiplication equation for the area shown by the rectangle, and find the perimeter of each rectangle.

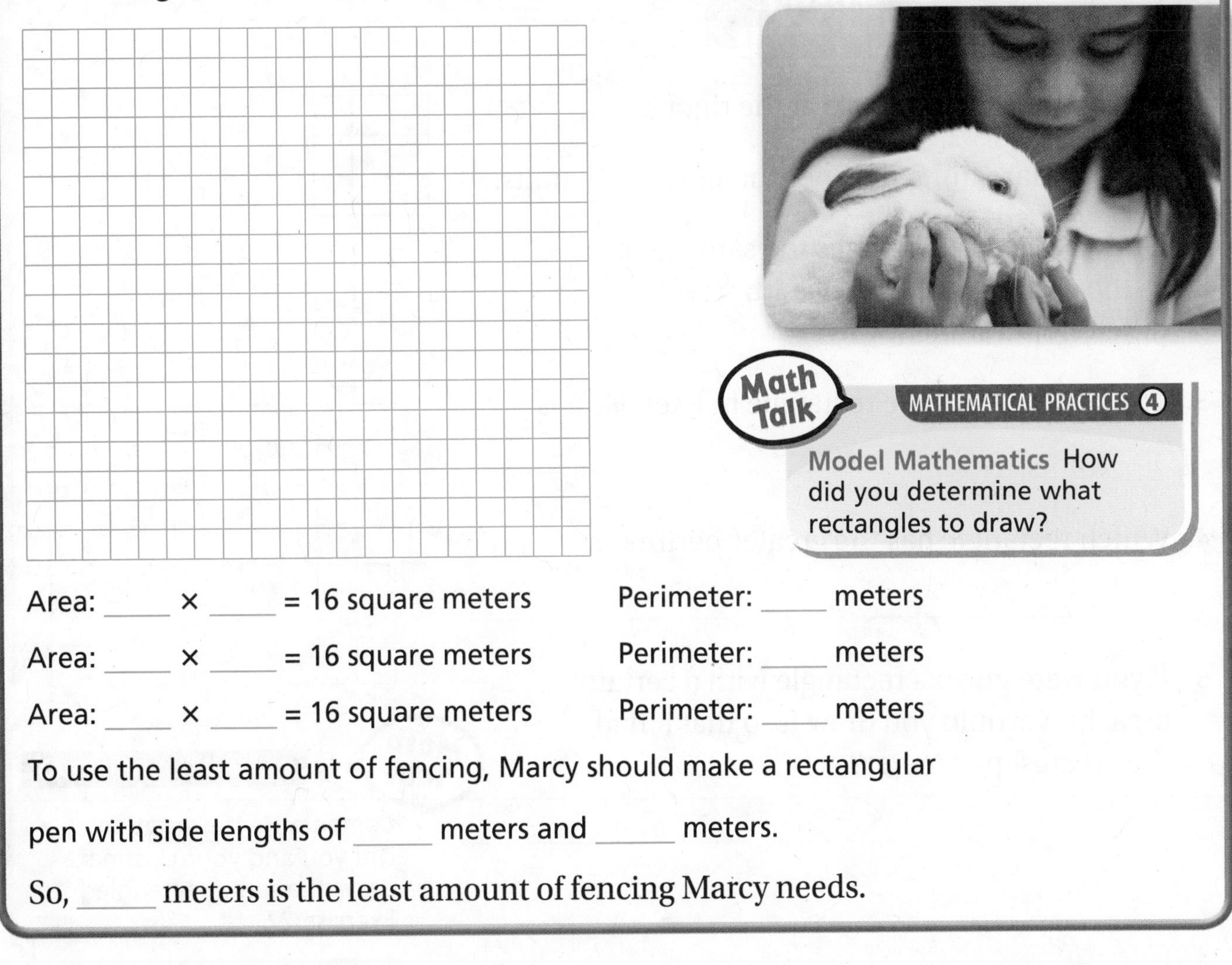

Math Talk MATHEMATICAL PRACTICES 4

Model Mathematics How did you determine what rectangles to draw?

Area: _____ × _____ = 16 square meters Perimeter: _____ meters

Area: _____ × _____ = 16 square meters Perimeter: _____ meters

Area: _____ × _____ = 16 square meters Perimeter: _____ meters

To use the least amount of fencing, Marcy should make a rectangular pen with side lengths of _____ meters and _____ meters.

So, _____ meters is the least amount of fencing Marcy needs.

Try This!

Draw three rectangles that have an area of 18 square units on the grid. Find the perimeter of each rectangle. Shade the rectangle that has the greatest perimeter.

Share and Show

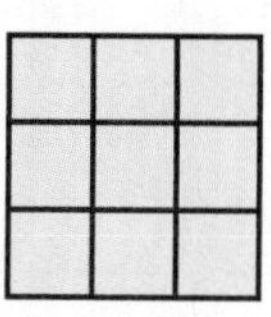

1. The area of the rectangle at the right is

   _____ square units. The perimeter is _____ units.

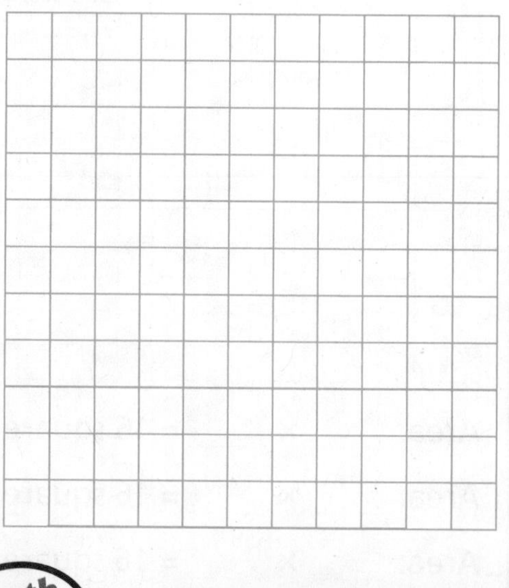

2. Draw a rectangle that has the same area as the rectangle in Exercise 1 but with a different perimeter.

3. The perimeter of the rectangle in Exercise 2 is

   _____ units.

✔ 4. Which rectangle has the greater perimeter?

   ____

5. If you were given a rectangle with a certain area, how would you draw it so that it had the greatest perimeter?

   ____

MATHEMATICAL PRACTICES 3

Compare Representations
Did you and your classmate draw the same rectangle for Exercise 2?

Name ______________________________

Find the perimeter and the area. Tell which rectangle has a greater perimeter.

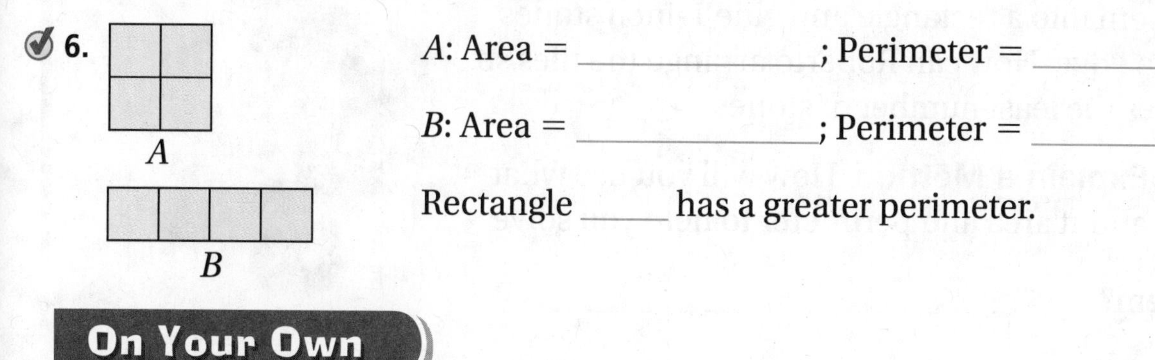

6. *A*: Area = ______________; Perimeter = ________

B: Area = ______________; Perimeter = ________

Rectangle _____ has a greater perimeter.

On Your Own

Find the perimeter and the area. Tell which rectangle has a greater perimeter.

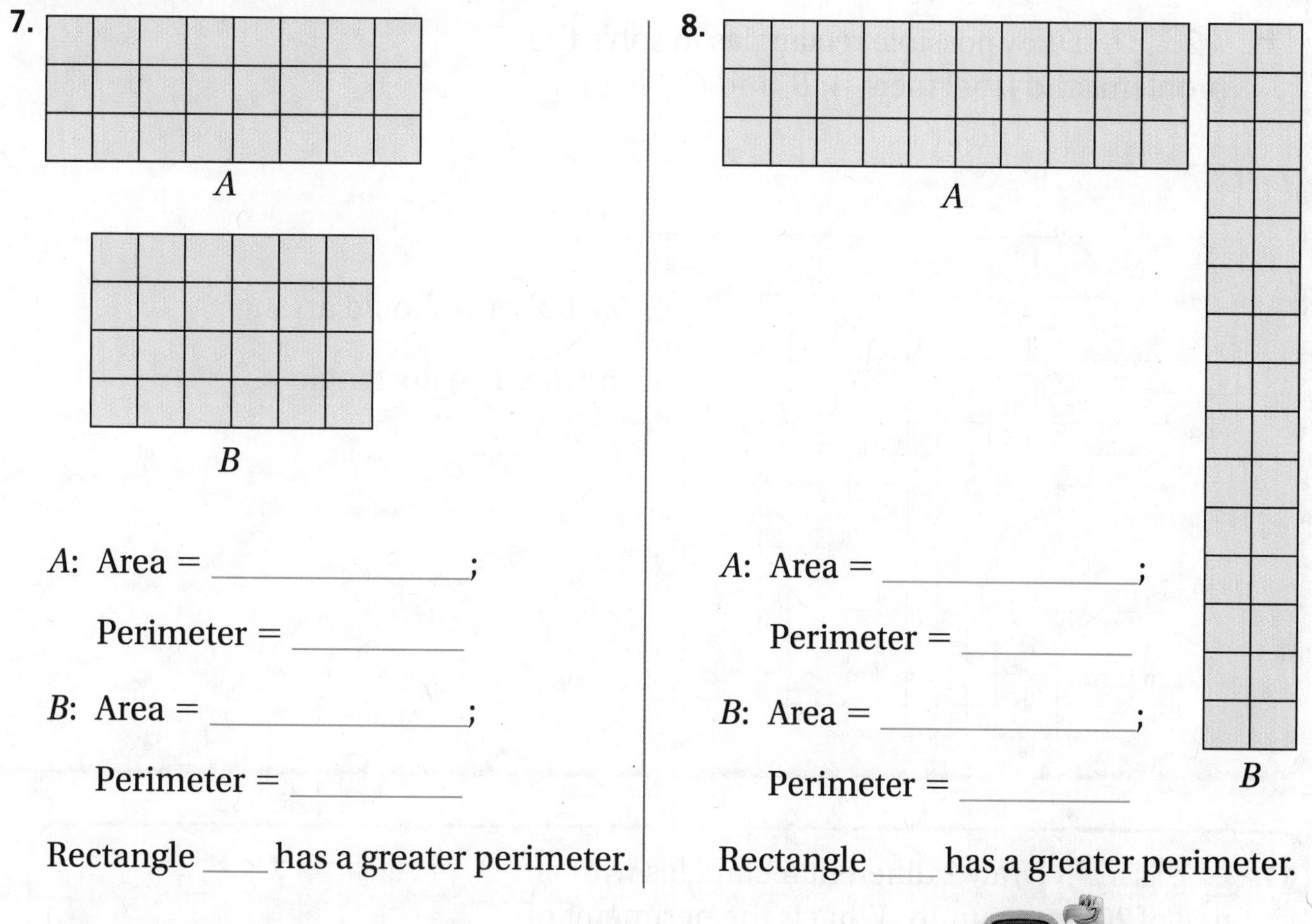

7. *A*: Area = ______________;

Perimeter = __________

B: Area = ______________;

Perimeter = __________

Rectangle ____ has a greater perimeter.

8. *A*: Area = ______________;

Perimeter = __________

B: Area = ______________;

Perimeter = __________

Rectangle ____ has a greater perimeter.

9. THINKSMARTER **Sense or Nonsense?** Dora says that of all the possible rectangles with the same area, the rectangle with the largest perimeter will have two side lengths that are 1 unit. Does her statement make sense? Explain.

Unlock the Problem Real World

10. Roberto has 12 tiles. Each tile is 1 square inch. He will arrange them into a rectangle and glue 1-inch stones around the edge. How can Roberto arrange the tiles so that he uses the least number of stones?

a. MATHEMATICAL PRACTICE 6 **Explain a Method** How will you use what you know about area and perimeter to help you solve the problem? ____________________

____

____

b. GO DEEPER Draw possible rectangles to solve the problem, and label them *A*, *B*, and *C*.

____

____

c. So, Roberto should arrange the tiles like Rectangle ____.

11. THINK SMARTER Draw 2 different rectangles with an area of 20 square units. What is the perimeter of each rectangle you drew?

Area = 20 square units

Perimeter = ____ units

Perimeter = ____ units

Name ______________________

Same Area, Different Perimeters

COMMON CORE STANDARD—3.MD.D.8
Geometric measurement: recognize perimeter as an attribute of plane figures and distinguish between linear and area measures.

Find the perimeter and the area. Tell which rectangle has a greater perimeter.

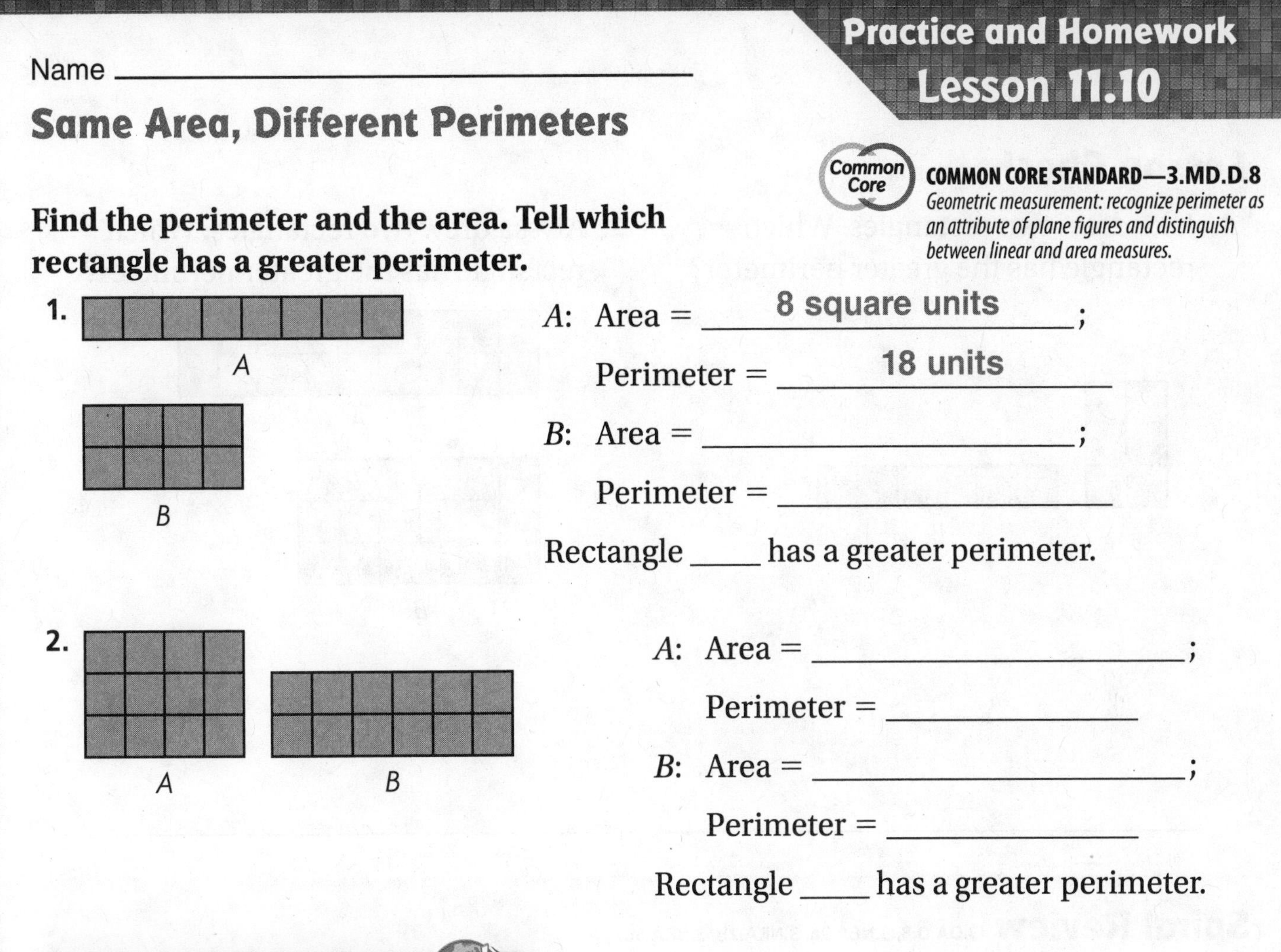

1. *A*: Area = 8 square units;
   Perimeter = 18 units
   B: Area = ______________;
   Perimeter = ______________
   Rectangle ____ has a greater perimeter.

2. *A*: Area = ______________;
   Perimeter = ______________
   B: Area = ______________;
   Perimeter = ______________
   Rectangle ____ has a greater perimeter.

Problem Solving (Real World)

Use the tile designs for 3–4.

Beth's Tile Designs

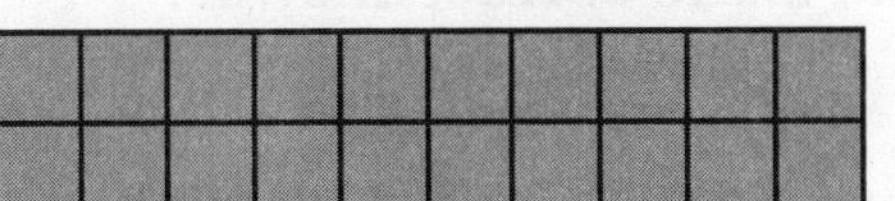

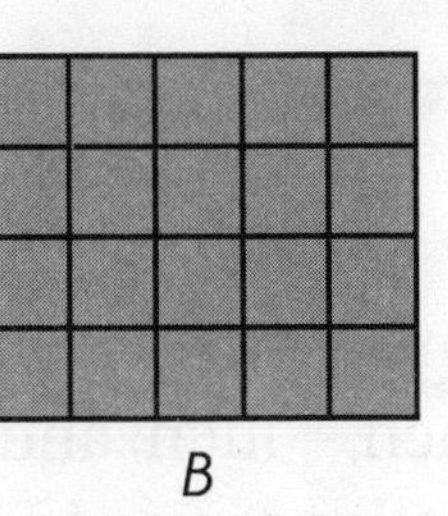

3. Compare the areas of Design A and Design B.

   ____

   ____

4. Compare the perimeters. Which design has the greater perimeter?

   ____

5. WRITE Math Draw two rectangles with different perimeters but the same area.

   ____

   ____

Lesson Check (3.MD.D.8)

1. Jake drew two rectangles. Which rectangle has the greater perimeter?

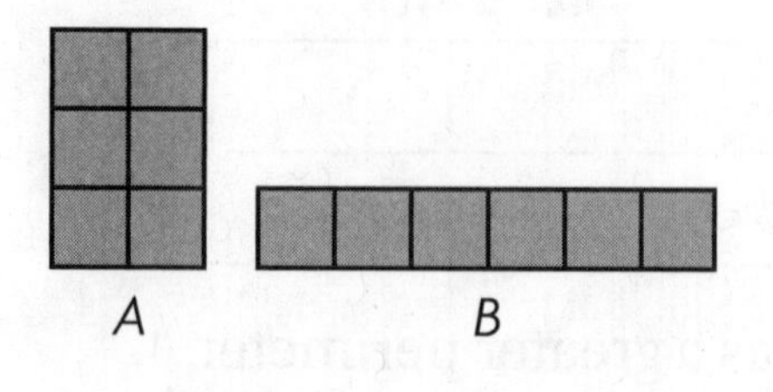

2. Alyssa drew two rectangles. Which rectangle has the greater perimeter?

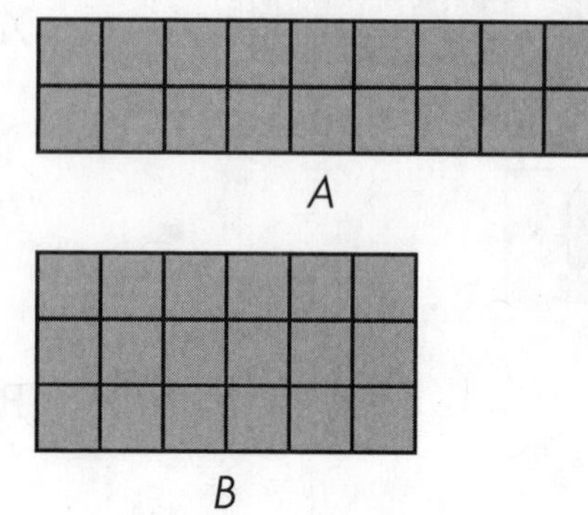

Spiral Review (3.OA.D.8, 3.NF.A.2a, 3.NF.A.2b, 3.NF.A.3b)

3. Marsha was asked to find the value of $8 - 3 \times 2$. She wrote a wrong answer. What is the correct answer?

4. What fraction names the point on the number line?

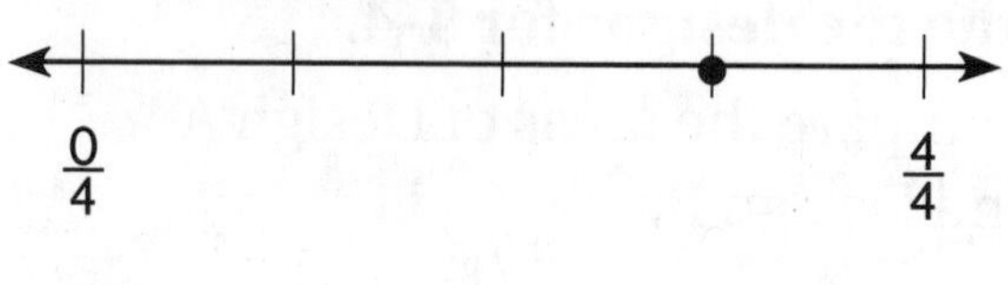

5. Kyle drew three line segments with these lengths: $\frac{2}{4}$ inch, $\frac{2}{3}$ inch, and $\frac{2}{6}$ inch. List the fractions in order from least to greatest.

6. On Monday, $\frac{3}{8}$ inch of snow fell. On Tuesday, $\frac{5}{8}$ inch of snow fell. Write a statement that correctly compares the snow amounts.

Name ____________________

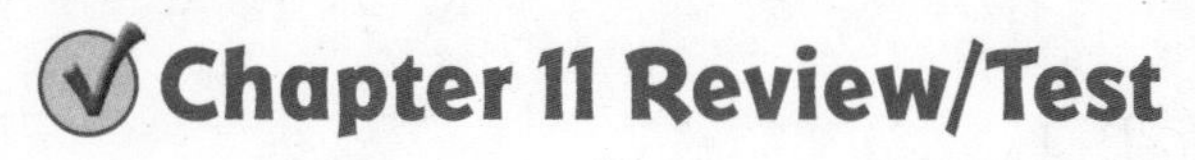

Chapter 11 Review/Test

1. Find the perimeter of each figure on the grid. Identify the figures that have a perimeter of 14 units. Mark all that apply.

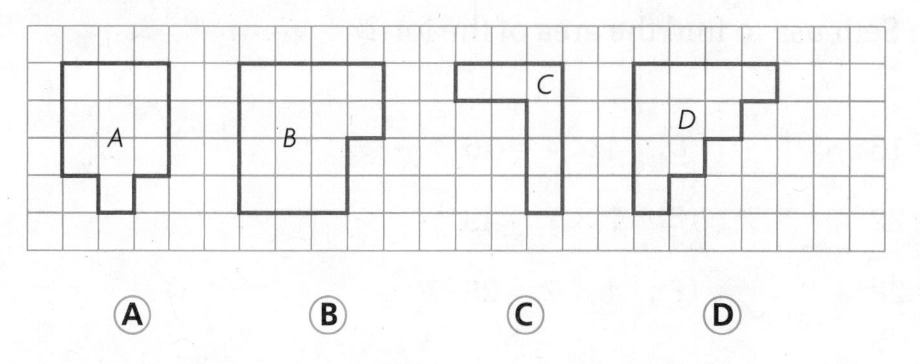

Ⓐ Ⓑ Ⓒ Ⓓ

2. Kim wants to put trim around a picture she drew. How many centimeters of trim does Kim need for the perimeter of the picture?

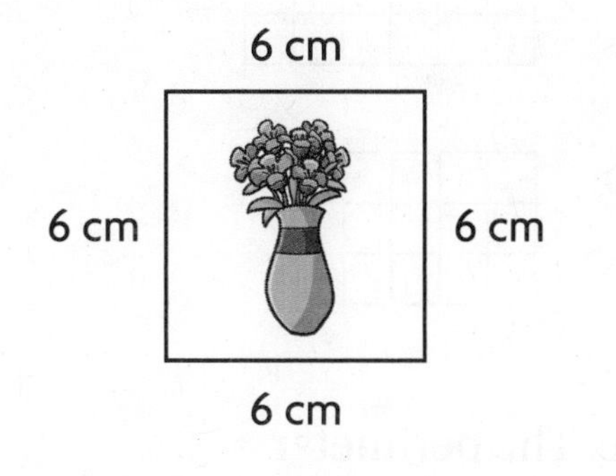

_____ centimeters

3. Sophia drew this rectangle on dot paper. What is the area of the rectangle?

_____ square units

Assessment Options
Chapter Test

4. The drawing shows Seth's plan for a fort in his backyard. Each unit square is 1 square foot.

Which equations can Seth use to find the area of the fort? Mark all that apply.

Ⓐ $4 + 4 + 4 + 4 = 16$

Ⓑ $7 + 4 + 7 + 4 = 22$

Ⓒ $7 + 7 + 7 + 7 = 28$

Ⓓ $4 \times 4 = 16$

Ⓔ $7 \times 7 = 49$

Ⓕ $4 \times 7 = 28$

5. Which rectangle has a number of square units for its area equal to the number of units of its perimeter?

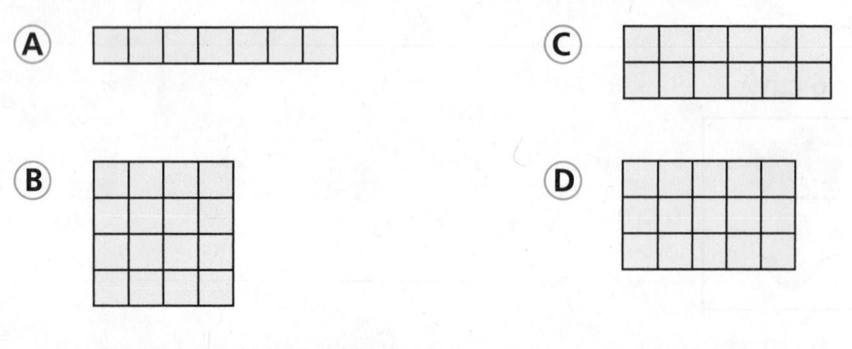

6. Vanessa uses a ruler to draw a square. The perimeter of the square is 12 centimeters. Select a number to complete the sentence.

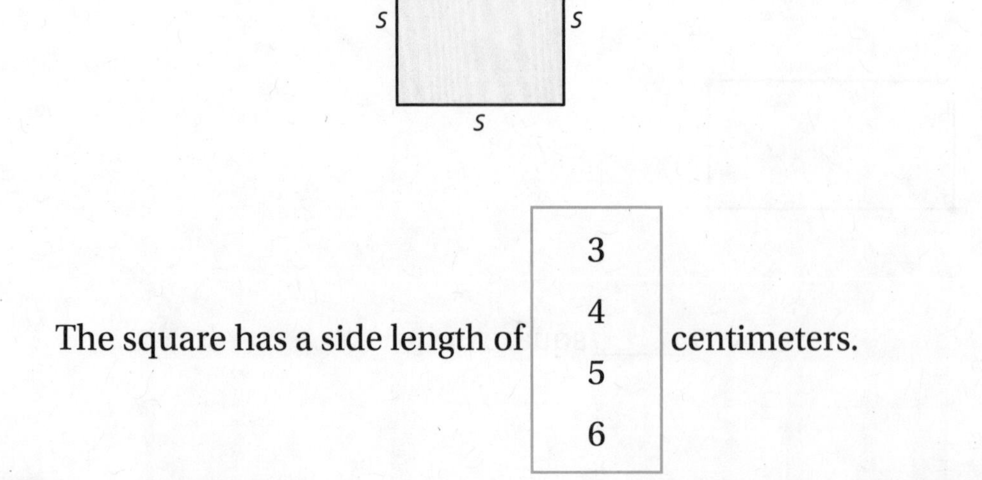

The square has a side length of [3 / 4 / 5 / 6] centimeters.

Name ______________________________

7. Tomas drew two rectangles on grid paper.

Circle the words that make the sentence true.

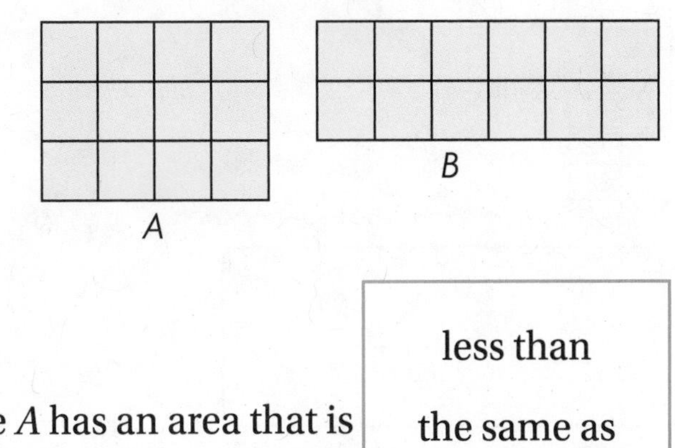

Rectangle *A* has an area that is

less than the same as greater than

the area of Rectangle *B*, and a perimeter that is

less than the same as greater than

the perimeter of Rectangle *B*.

8. Yuji drew this figure on grid paper. What is the perimeter of the figure?

_____ units

9. What is the area of the figure shown? Each unit square is 1 square meter.

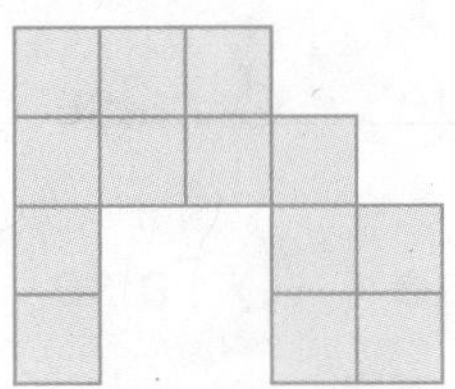

_____ square meters

10. Shawn drew a rectangle that was 2 units wide and 6 units long. Draw a different rectangle that has the same perimeter but a different area.

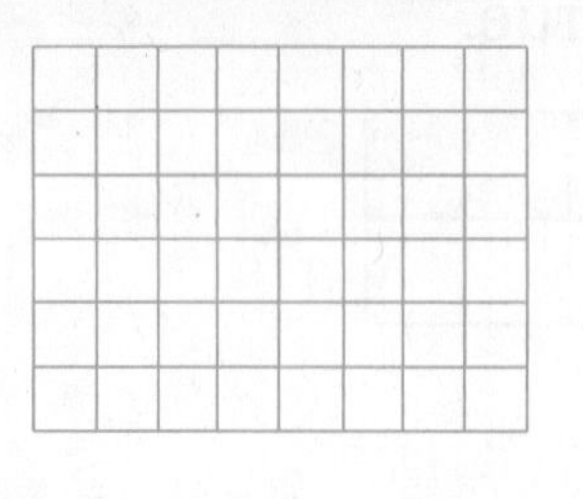

11. Mrs. Rios put a wallpaper border around the room shown below. She used 72 feet of wallpaper border.

What is the unknown side length? Show your work.

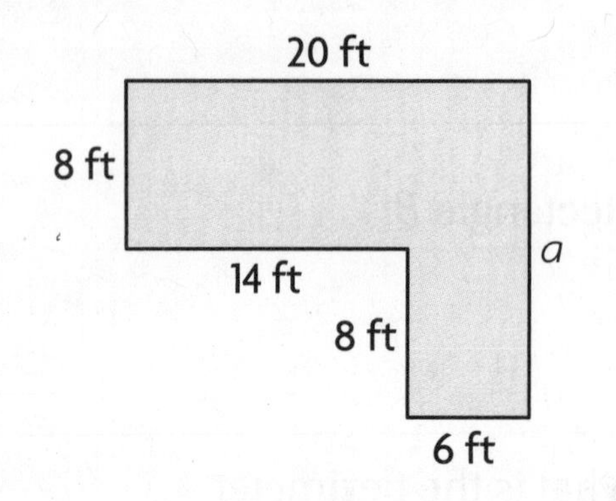

______ feet

12. Elizabeth has two gardens in her yard. The first garden is 6 feet long and 4 feet wide. The second garden is double the width of the first garden. The area of the second garden is double the area of the first garden. For numbers 12a–12d, select True or False.

12a. The area of the first garden is 24 square feet. ○ True ○ False

12b. The area of the second garden is 48 square feet. ○ True ○ False

12c. The length of the second garden is 12 feet. ○ True ○ False

12d. The length of the second garden is 6 feet. ○ True ○ False

Name ______________________________

13. Marcus bought some postcards. Each postcard had a perimeter of 16 inches. Which could be one of the postcards Marcus bought? Mark all that apply.

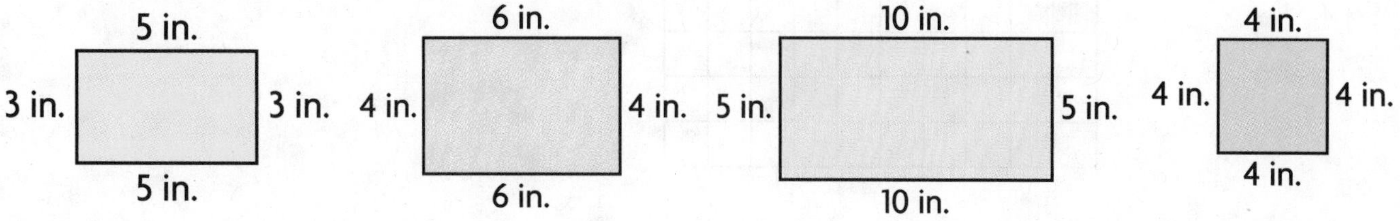

Ⓐ Ⓑ Ⓒ Ⓓ

Personal Math Trainer

14. THINK SMARTER+ Anthony wants to make two different rectangular flowerbeds, each with an area of 24 square feet. He will build a wooden frame around each flowerbed. The flowerbeds will have side lengths that are whole numbers.

Part A

Each unit square on the grid below is 1 square foot. Draw two possible flowerbeds. Label each with a letter.

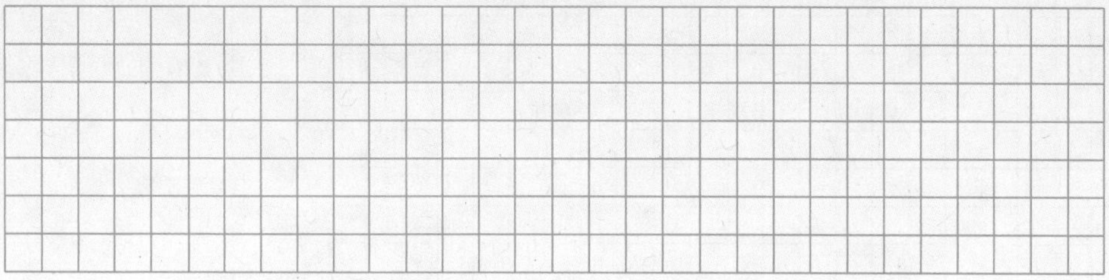

Part B

Which of the flowerbeds will take more wood to frame? Explain how you know.

____

____

____

15. Keisha draws a sketch of her living room on grid paper. Each unit square is 1 square meter. Write and solve a multiplication equation that can be used to find the area of the living room in square meters.

_____ square meters

16. GO DEEPER Mr. Wicks designs houses. He uses grid paper to plan a new house design. The kitchen will have an area between 70 square feet and 85 square feet. The pantry will have an area between 4 square feet and 15 square feet. Draw and label a diagram to show what Mr. Wicks could design. Explain how to find the total area.

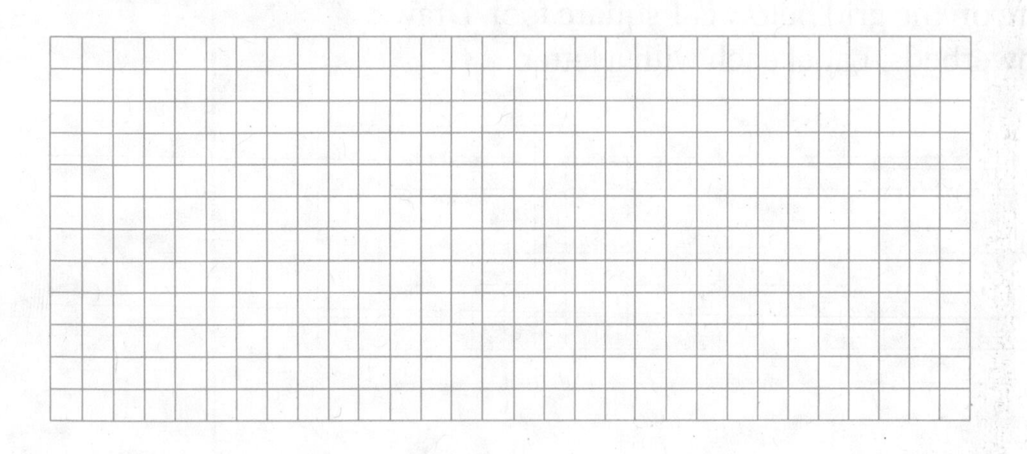